SOTHIS PRESS

Sothis Press aims to completely reimagine the public domain book with an emphasis on universal access to captivatingly designed print editions. We believe that technology will enable every single book ever published to be perpetually accessible in beautiful print form. Sothis is beauty in ubiquity.

www.sothis.press
1 Waterhouse Square, London, ECIN 2ST

Railway Engineering

Or, Field Work Preparatory to the Construction Railways; Containing the Original and Most Approved Methods of Laying Out Railway Curves, and of Setting Out the Widths of Cuttings and Embankments, Etc;; A General Table for the Calculation of Earthworks Of

T. Baker

SOTHIS PRESS

ISBN 978-1-0365-1147-0

This edition is published by
Sothis Press © 2023. The company
is registered in England and Wales.
The Company Number is 14900389.

RAILWAY ENGINEERING;

OR,

FIELD WORK

PREPARATORY TO THE CONSTRUCTION OF RAILWAYS:

CONTAINING

THE ORIGINAL AND MOST APPROVED METHODS OF LAYING OUT RAILWAY CURVES, AND OF SETTING OUT THE WIDTHS OF THE CUTTINGS AND EMBANKMENTS, ETC.;

A General Table

FOR THE

CALCULATION OF EARTHWORKS

OF

RAILWAY CANALS, &c.

WITH TWO AUXILIARY TABLES;

ALSO,

TUNNELLING, AND INVESTIGATION OF THE FORMULA FOR THE SUPER-ELEVATION OF THE EXTERIOR RAIL IN CURVES.

BY

T. BAKER, C.E.

AUTHOR OF

"A System of Surveying by the Theodolite," and "Railway Engineering," in the Ninth Edition of Nesbit's Surveying; "Integration of Differentials;" &c. &c.

Second Edition.

LONDON:

LONGMANS, GREEN, READER, & DYER.

1870.

TO

PETER BARLOW, ESQ.,

F.R.S., M. INST. C.E., ETC. ETC.

LATE PROFESSOR OF MATHEMATICS IN THE ROYAL MILITARY ACADEMY, WOOLWICH,

AND ONE OF THE COMMISSIONERS OF THE RAILWAY GAUGES,

AS

A MARK OF ESTEEM

FOR HIS DISINTERESTED KINDNESS,

AND

IN ADMIRATION OF HIS TALENTS

AS AN ORIGINAL MATHEMATICIAN AND PHILOSOPHER,

This Work is Dedicated,

WITH THE HONOUR OF HIS PERMISSION,

BY HIS MOST DEVOTED SERVANT,

THE AUTHOR.

PREFACE TO THE FIRST EDITION.

THE matter contained in the following pages, with the exception of a few additions and improvements, was drawn up about twenty-five years ago for private use; its publication, at the present time, having been delayed at least two years, for reasons, which it will be unnecessary here to state.

The chief part of SECTION I., On the Different Methods of Laying out Railway Curves on the ground, was communicated to my friend the late Professor Leybourn, of the R. M. College, Sandhurst, in 1824, and afterwards published by him in the "Gentleman's Diary;" these Methods being adapted to all cases that can occur in practice, as well as to the different degrees of scientific skill possessed by surveyors and engineers. The Constructions, Formulæ, and Methods of Laying out Compound, Serpentine, and Deviation Curves, with Tables of Offsets for Railway Curves, were also placed in his hands; but his death prevented their publication.

As these Methods of Laying out Railway Curves, especially two of them, have been generally adopted in practice, from the commencement of the Stockton and Darlington Railway to the present time, and as several other gentlemen, during the last two or three years, have published on the same subject, most of whom have given either all or part of my methods, I think it right to claim the Invention of them.—(See Remarks on the Invention of the Methods of laying out Railway Curves, p. 29.)

The gentlemen that have published on this subject are, Law, Castle, Rankine, and Hill (see page 31). I have lately seen similar publications by Heald, Brodie, Gardner, May, and others. Almost all these authors give their formulæ and rules for laying out curves without investigation; some of them have given an unnecessary profusion of formulæ, which involve the subject they pretend to explain, in such a degree of obscurity as must be very perplexing to students, and repulsive to those engineers who have been accustomed to use my methods, *in which I anticipated, above twenty years ago, all that has since been done by these authors, at least as far as real practical utility is concerned.*

In SECTION II., I have given Methods of setting out the widths of ground for Railway Cuttings, on horizontal, laterally sloping, and uneven ground, with original Formulæ connected therewith; superseding, in most cases, the unscientific methods of approximation given for the same purpose used by many engineers. This Section concludes with methods of calculating the quantity of land required for projected and intended Railways.

In SECTION III., I have treated extensively on finding the Contents of Railway Cuttings, both for preliminary Estimates from the depths, and for actual Cuttings from Sectional Areas, by means of a General and two Auxiliary Tables, on one folding Sheet at the end of the work.—*The Method here given of finding the Contents of Cuttings from Sectional Areas is, as far as I know, the only one mathematically correct, yet published,* having been prepared for private use many years ago.

The authors who have published Tables on this subject are, Sir John M'Neill, Mr Bidder, Mr Bashforth, Messrs Sibley and Rutherford, Mr Huntingdon, Mr Law, and others; some of whose Tables are voluminous, and most of them well adapted for finding the Contents of Cuttings for preliminary Estimates; but none of them are accompanied with directions for finding the Contents from Sectional Areas, which is the most important part of the use of such Tables, except Mr Bashforth's; *but his method of applying them is erroneous.*—(See pages 41 and 54.) This Section contains Investigations of the Methods of Constructing the Tables, and of using them for finding the Contents of Cuttings; and concludes with investigating and pointing out the errors of Mr Bashforth and others.

In SECTION IV., I have given Methods of setting out the Earthwork of Tunnels, with a Table of the Dimensions of several existing Tunnels.

In SECTION V., I have given the Investigation and Application of the celebrated Formulæ for the Superelevation of the Exterior Rail in Railway Curves, which is, I believe, due to *Pambour.*

The matter contained in this Work is drawn up in a manner which, I trust, will be easily understood by any surveyor or engineer; Examples, wrought out at length, being given in every case, the Mathematical Investigations of the Problems, &c., being kept separate from the practical parts of the Work, the whole of which, except a few trifling alterations, form part of a system of Railway Engineering, which I added to the Ninth Edition of Nesbit's Surveying, lately published.

In conclusion, I can truly say that the whole of the matter contained in this small volume, with some few trifling exceptions, was originally drawn up by me; the most essential parts of which have either been published or communicated to my friends for periods varying from ten to twenty-five years; no author, on the subjects here treated of, having preceded me, whose results appeared to be of a character adapted to this work, excepting one of *Pambour's*, as already noticed; and a small part of the investigation of the rules for finding the Contents of Earthworks which has been given, as is well known, by various authors.

T. BAKER.

Vauxhall Street, London, 1848.

CONTENTS.

SECTION I.—RAILWAY CURVES.

SECTION II.—RAILWAY EARTH WORKS.

SECTION III.—CONTENTS OF CUTTINGS.

SECTION IV.

SECTION V.

SECTION VI.

RAILWAY ENGINEERING;

OR,

FIELD WORK

PREPARATORY TO THE CONSTRUCTION OF RAILWAYS.

SECTION I.

RAILWAY CURVES.

(1.) *On the use of curves in railways in general.*

The use of curves in railways is absolutely necessary on account of the natural unevenness of the earth's surface, it being desirable to attain the nearest possible practical level by avoiding hills, crags, mountains, &c., and by winding round them by means of curves. Curves are also equally necessary in avoiding other natural and artificial obstructions, in many cases not materially affecting the level of the line, as lakes, swamps, the windings of sea-coasts, and of rivers; also cities, towns, villages, parks, pleasure-grounds, &c. In this manner a great saving is effected in the expense of extensive excavations, embankments, tunnels, viaducts, &c., as well as the expense of the severance of valuable property, which would otherwise be required. Besides, it is frequently desirable to make a winding railway, in order that it may embrace in its route some important city, town, harbour, &c., or make a junction with another railway. Straight lines in railways are, however, much to be preferred to curves, and are therefore adopted as far as possible; they are first set out as bases for further operations; and every curve is subordinate to two straight portions of the line, which are tangents to it at its commencement and termination. The curves adopted in practice are always arcs of circles. Sometimes two, three, or more consecutive circular arcs, having a common tangent at their point or points of junction, are joined together, as in the case of the compound curve; and sometimes two circular arcs are connected, having their convexities turned different ways, with a common tangent at their point of junction, as in the case of the serpentine, or S curve.

NOTE.—Other curves beside the circular arc might sometimes with advantage be adopted as the curves of railways; but their construction is attended with difficulty.

(2.) *On mechanical railway curves, or curve-rulers.*

Mechanical railway curves, sometimes called curve-rulers, are a series of segments of circles made of hard wood, as box or mahogany, or strong paste-board, with their radii in inches marked on them. These curves

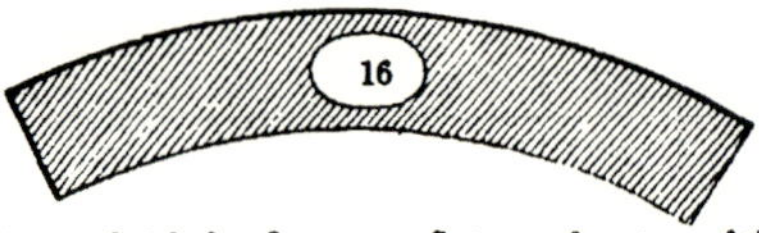

commonly begin with a radius of $2\frac{1}{2}$ inches, and terminate with one of 160 inches, or upwards; the smallest radii usually increased by $\frac{1}{2}$ inch up to 10 inches; then by inches up to 20 inches; afterwards by 2 inches up to 80, including all numbers ending in 5; lastly, by 5 inches, till near the end of the series, when the radii increase by 10 inches.

The annexed figure represents the railway curve-ruler of 16 inches radius. If the scale of plan, to which this curve-ruler is applied be 5 chains to an inch, it will represent a curve or circular arc of $16 \times 5 = 80$ chains $= 1$ mile radius. If the scale of the plan be 12 chains to an inch, it will represent an arc of $16 \times 12 = 192$ chains $=$ 2 miles and $3\frac{1}{5}$ furlongs radius; and similarly for other scales.

(3.) *The limit of the radii of railway curves.*

As a general rule the curves of a railway should be kept of as large a radius as possible consistently with the economical laying-out of the line. One mile radius or upwards is quite unobjectionable; but half-a-mile, or even less, will be sanctioned by Parliament in difficult cases, particularly near large stations, where the speed is usually slackened. In all sharp curves the outer rail requires to be raised above the level of the inner one, to counteract the effect of the centrifugal force. (See Formulæ in Section VII.)

PROBLEM I.

CASE I.

To determine the radius of a railway curve mechanically, that is, by the curve-rulers; the positions of the straight or tangential portions of the railway being given.

This problem admits of an indefinite number of solutions, but the curve adopted is generally that which falls on ground presenting the fewest natural or artificial obstructions, provided its radius be not too great. (See Art. 3.)

Let AB, CD be two straight portions of a railway, the positions of which are given by an accurate plan; and which, when prolonged, meet at T. Apply several of the series of curve-rulers (see Art. 2) to touch the line AB, without cutting it, at the points B, B′, B″, &c., and also to touch CD, in like manner, at the points C, C′, C″, &c.; and let BC, BC′, BC″. &c., be the curves thus obtained. Then whichsoever of these curves may appear, on a careful consideration, to be least obstructed by hills, lakes, rivers, buildings, &c., must be adopted as part of the line of railway, provided its radius be equal to, or exceed, the required limit. The radius is determined by multiplying the number on the curve-ruler by the number of chains per inch on the scale of the plan.

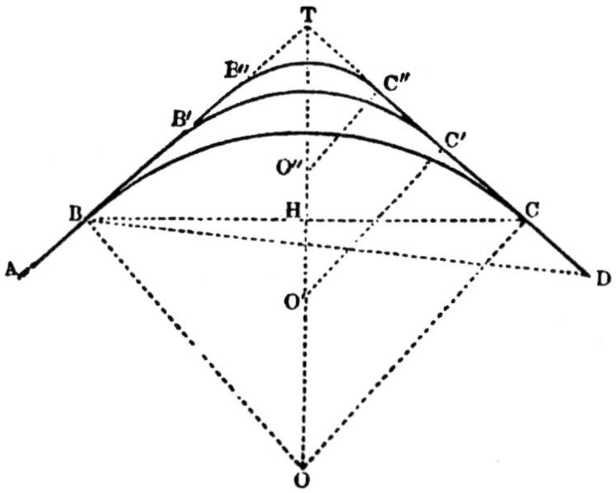

Example. If the curve BC be drawn by the curve-ruler, numbered 22, and be adopted as part of the line, and the scale of the plan be 5 chains to an inch, required the radius of the curve.

$22 \times 5 = 110$ chains $= 1$ mile 3 fur., the radius required.

NOTE.—By this method the radii of railway curves are commonly found in practice; but since a circular arc of large radius apparently coincides with its

tangent for a considerable distance, it is difficult, in this manner, to determine the exact starting-points of the curve; besides, if the plan be not strictly accurate, as is too often the case, the positions of the tangents cannot be said to be given, and therefore this method should only be employed in roughly determining the radius and starting-points of the curve, so that it may fall, in its progress, on the most favourable ground.

CASE II.

To determine the radius of a railway curve geometrically, the starting-point B *being given.*

Prolong AB, DC till they meet at T; bisect the angle ATD by the line TO; and at B draw BO perpendicular to AT, meeting TO in O. Then O is the centre, and BO the required radius of the curve BC.

Or,

Make TC = TB; and draw the perpendiculars BO, CO, meeting in O. Then BO, or CO, is the radius of the curve.

In the same manner, the other radii C′O′, C″O″, &c., of the curves B′C′, B″C″, &c., may be found.

CASE III.

To find the radius of the curve by taking dimensions on the ground, the map being known to be inaccurate, and the starting-point B *being given.*

Range the tangents AB, DC till they meet at T; measure BT; then measure on TD till TC = BT; also measure BC. Then $BO = \frac{1}{2} BC . BT \div \sqrt{BT^2 - \frac{1}{4} BC^2}$ * = radius required.

Ex. Let BT = 82·50 chains, and BC = 132·00 chains. Then BO = $\frac{1}{2}(132 \times 82{\cdot}5) \div \sqrt{(82{\cdot}5)^2 - \frac{1}{4}(132)^2} = 5445 \div 49{\cdot}5 = 110$ chains, the radius required.

Or by Trigonometry.

Range the tangents AB, DC till they meet at T, as before, then measure BT, and take the angle BTC, $\frac{1}{2}$ of which is the ∠BTO; then in the triangle BTO, right-angled at B, are given BT and the ∠BTO to find BO; whence by trigonometry,

cot. ∠BTO : rad. : : TC : BO, the radius required.

Ex. Let BT = 82·50 chains, and the angle BTC = 106° 16′.

Then cot. $\frac{1}{2}$ ∠BTC = cot. ∠BTO, 53° 8′ . .	9·87501
: rad. or sin. 90°	10·00000
: : BT = 82·50 chains	1·91645
: BO = 110 chains, the required radius . .	2·04144

* *Demonstration.*—In the Δ BTH, $TH = \sqrt{BT^2 - BH^2} = \sqrt{BT^2 - \frac{1}{4} BC^2}$, and by the similar Δs BHT, BHO, $\sqrt{BT^2 - \frac{1}{4} BC^2}$ (= TH) : BT : : $\frac{1}{2}$ BC (= BH) : BO = $\frac{1}{2}$ BC : BT $\div \sqrt{BT - \frac{1}{4} BC^2}$. *Q. E. D.*

Case IV.

To find the radius of the curve, when the plan is known to be inaccurate, and the tangents AB, DC *cannot be prolonged to meet at* T *on account of obstructions, the starting-point* B *of the curve being given.*

From B measure a line BD, on the most convenient ground, to meet CD at D, taking the angles TBD, TDB; the sum of which taken from 180° gives the ∠ BTD, whence by trigonometry,

sin. ∠ BTD : BD : : sin. ∠ TDB : BT.

Now in the right-angled triangle BTO are given BT and the angle BTO = ½ ∠BTD, to find the radius BO, which may be done as in Case III.

PROBLEM II.

To lay out on the ground a railway curve by the common method.

Let AB, CD be two straight or tangential portions of a railway, the points BC being required to be joined by a circular curve BC, to which AB, CD shall be tangents at the points B and C; and let BO be the radius of the curve, which is supposed to be determined by one or

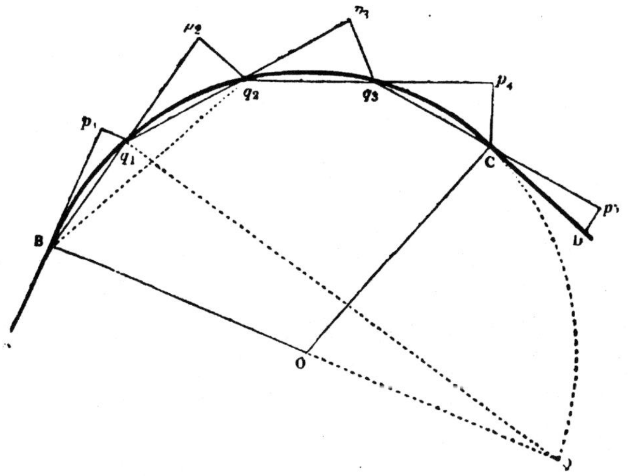

other of the cases in Problem I. accordingly as the map is found to be accurate or inaccurate. Put the radius BO $= r$; measure on the tangent AB prolonged the distance Bp_1 = 1 chain, as is usual in practice; set off $p_1q = \frac{1}{2r}$* at right angles to Bp_1; then q_1 is the first point in the curve. Through Bq_1 measure the right line Bq_1p_2 = twice Bp_1 = 2 chains, and set off $p_2q_2 = \frac{1}{r}$ = twice p_1q_1 at right angles to Bq_1p_2; then q_2 is the second point in the curve, repeating the last operation till the curve reach the point C. Lastly, q_3Cp being measured = 2 chains the last offset p_5D will be found = the first offset p_1q_1, or half the preceding one p_4C, if the work be right. See the following notes and example.

NOTE 1. Only a small part of the operation for a curve of great length is shown in the figure; but as the whole of the work, except the first and last offsets, is alike, to show more would be unnecessary.

2. The values of $\frac{1}{2r}$ for all radii, from 15 chains to 540, are given in Table No. 4, at the end of the work.

Ex. Let BO = r = 80 chains = 1 mile; then $\frac{1}{2r} = \frac{1}{160}$ of a chain, which being multiplied by 792, the number of inches in 1 chain, gives $\frac{1 \times 792}{160}$ = 4·95 inches, the first offset $p\,q_1$; whence 4·95 × 2 = 9·9 inches, the second offset p_2q_2.

Or by Table No. 4, opposite 80, in the column marked Radius, &c., stands 4·95 inches, in the column marked Offsets, &c., which is the value of p_1q_1, as above.

3. By this method the greater part of British, as well as foreign, railway curves have been laid out, having been invented by Mr T. Baker, C.E., at a very early period, when the Stockton and Darlington railway was laid out. It was eagerly adopted by railway surveyors, as it involves very little calculation, and does not require the use of an angular instrument. It is, however, defective in practice, on account of its requiring the coupling together of so very many short lines, as small errors will unavoidably creep in and multiply (and more especially so if the ground be rough), so that the curve has frequently to be retraced several times

---

* *Demonstration.*—Complete the semicircle BCQ, prolong BO to Q, and join q_1Q, Bq_2. Then because Bp_1, Bp_2, Bq_2, are always, in practice, so very small compared with BQ that they nearly coincide with the curve, and consequently Bp_1 is approximately = Bq_1 = q_1p_2 = q_1q_2 = ½Bq_2; ∴ by the nature of the circle the Δs Bq_1Q, Bp_1q_1, Bp_2q_2, are similar, and

$$\text{BQ} : \text{B}q_1 :: \text{B}q_1 : p_1q_1 = \frac{\text{B}q_1^2}{\text{BQ}} = \frac{\text{B}p_1^2}{\text{BQ}} = \frac{\delta^2}{2r}.$$

$$\text{also} \quad : \quad :: \text{B}q_2 : p_2q_2 = \text{B}q_1 \times \text{B}q_2 = \frac{2\text{B}p_1^2}{\text{BQ}} = \frac{\delta^2}{r}.$$

BO = ½BQ being put = r, and Bp_1 = Bq_1 = &c. = δ; and when δ = 1 chain, as in practice, $p_1q_1 = \frac{1}{2r}$, and $p_2q_2 = p_3q_3$ = &c. = $\frac{1}{r}$. *Q. E. D.*

before it can be got right, especially if it be a long one. *This defect in the above method induced the author to prepare three other and more accurate methods, which are given in the following Problems.*

4. When the curve shall have been laid out with sufficient accuracy, the rods or quills that mark the ends of the offsets, $q_1 q_2$, &c., must be taken out and their places supplied by strong wooden stumps, about 16 or 18 inches in length, and $1\frac{1}{2}$ inch square, each end of the curve being marked with three stumps, or one large one with a cross or some other conspicuous mark on it. The straight portions of the line must also be marked with stumps at the end of every chain.

PROBLEM III.

To lay out a railway curve by ordinates or offsets from its tangents, no material obstructions being supposed to exist on the convex side of the curve, to prevent the use of the chain.

Case I.

When the length of the curve is less than $\frac{1}{4}$ of its radius.

Let BC be the curve, AB, CD, tangents thereto, which are ranged till they meet at D, and BO the radius of the curve, which must be correctly determined by one or other of the cases in Prob. I.

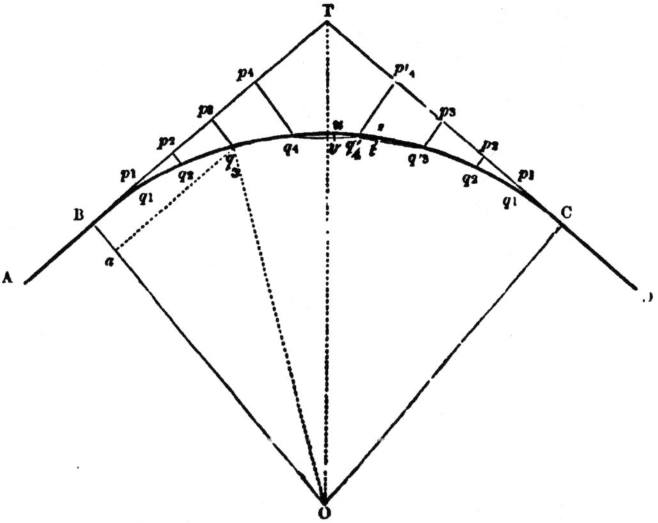

according to circumstances. Measure on BT the usual distance Bp_1 = 1 chain, and lay off the offset $p_1 q_1 = \frac{1}{2r}$ (r being the radius of the curve as before), then measure $p_1 p_2$ = 1 chain, and lay off

$p_2q_2 = \frac{2^2}{2r}$. The successive offsets at the end of every chain being $\frac{1}{2r}, \frac{2^2}{2r}, \frac{3^2}{2r}, \frac{4^2}{2r}$, &c.,* or, what amounts to the same thing, the 2d, 3d, 4th, &c., offsets are respectively 2^2, 3^2, 4^2, &c., times the first offset, or 4, 9, 16, &c., times the first one.

Having laid out the offsets in this manner, till the last one, which suppose to be p_4q_4 is either within, or very little more than, a chain from T, make $p'_4\text{T} = p_4\text{T}$, and lay out the same offsets in an inverted order, on TC, as were laid out on BT, that is, beginning with the greatest first and ending with the least.

Ex. Let the radius of the curve be 80 chains, then $p_1q_1 = \frac{1 \times 792}{2 \times 80}$ $= 4{\cdot}95$ inches; whence $p_2q_2 = 4 \times 4{\cdot}95 = 19{\cdot}8$ inches, $p_3q_3 = 9 \times 4{\cdot}95$ $= 44{\cdot}55$ inches $= 3$ ft. $8{\cdot}55$ in., &c., or the value of the first offset may be taken from the table at the end of the book.

To make the distances of the stumps equal.

As it can rarely happen in practice that the distance $q_4q'_4$ will be $= 1$ chain, when this is the case it will be better to set out the curve from C as was done from B; then, if the distance $q_4q'_4$ be less than 1 chain, suppose m links less, lay the chain from q'_4 to q'_3, and at the distance $q'\,t = m$, lay off the offsets $ts = \frac{m(1-m)}{2r}$, then s will be the point for the stump, instead of q'_4, thus making $q'_4s = 1$ chain. This operation must be repeated between q_3 and q_2, q_2 and q_1, &c., the last point to be changed necessarily falling on the tangent CD. If the distance $q_4q'_4$ be greater than 1 chain by m links, lay the chain

* *Demonstration.*—Draw the radius q_2O, and $q_2\,a \perp$ to OB. Put $\text{OB} = \text{O}q^2 = r$, and $q_2a = p_2\text{B} = \delta$; then $\text{O}a = \sqrt{r^2 - \delta^2}$, and $\text{B}a = p_2q_2 = \text{OB} - \text{O}a = r - \sqrt{r^2 \delta^2}$. Now, if δ be taken successively $= 1^2, 2^2, 3^2$, &c., chains, the values of p_1q_1, p_2q_2, p_3q_3, &c., will be respectively $r - \sqrt{r^2 - 1^2}$, $r - \sqrt{r^2 - 2^2}$, $r - \sqrt{r^2 - 3^2}$, &c. But as $1^2, 2^2, 3^2$, &c., are very small compared with r^2, within the limitation assigned to Case I. of this Problem, the values of $\sqrt{r^2 - 1^2}$, $\sqrt{r^2 - 2^2}$, &c., may be taken respectively $r - \frac{1^2}{2r}$, $r - \frac{2^2}{2r}$, &c., without material error, whence $p_1q_1 = r - \left(r - \frac{1^2}{2r}\right) = \frac{1^2}{2r}$, $p^2q^2 = r - \left(r - \frac{2^2}{2r}\right) = \frac{2^2}{2r}$, &c. But when the length of the curve exceeds $\frac{1}{4}$ of its radius, only 5 or 6 of the offsets ought to be taken in this manner, and the remainder, say, commencing with p_7q_7, ought to be taken $= r - \sqrt{r^2 - 7^2}$, &c.; although in a curve of 80 chains radius, the 10th offset by the contracted method exceeds the same offset by the correct method by only $\frac{1}{4}$ of a link out of $62\frac{1}{2}$ links. This difference, however, becomes gradually greater as the distance on the tangent approaches the 20th chain.

from q'_4 in the direction of q_4, and at the distance of $q'_4\,v = m$ links, lay off the offset $vu = \frac{(1+m)m}{2r}$, then u will be the place for the stump; next lay the chain from q_3 in the direction of u, and repeat the previous operation: this must be done to the end of the curve, the last mark to be changed falling, in this case, within the curve. The stumps may now be put down as pointed out in Note 4, Prob. II.

Ex. 1. When the distance $q_4q'_4$ is less than 1 chain by 40 links, the radius of the curve being 80 chains. Here $ts = \frac{(1-m)m}{2r} =$

$$\frac{(1-\cdot40)\times\cdot40}{160}\text{ ch.} = \frac{\cdot60\times\cdot40\times792}{160} = 1\cdot188\text{ in.} = 1\tfrac{1}{8}\text{ inch nearly.}$$

Ex. 2. When the distance $q\,q'_4$ is more than 1 chain by 40 links, and rad. = 80 chains. Here $vu = \frac{(1+m)m}{2r} = \frac{(1+\cdot40)\times 40}{160}\text{ch.}\times$

$$\frac{(1\cdot40\times\cdot40)\times792}{160} - 2\cdot742\text{ in.} = 2\tfrac{3}{4}\text{ inches nearly.}$$

CASE II.

When the curve is any required length.

As it is inconvenient in practice to have the offsets greatly to exceed two chains in length; if, therefore, the curve be a long one, the offsets may be confined within proper limits by dividing the curve

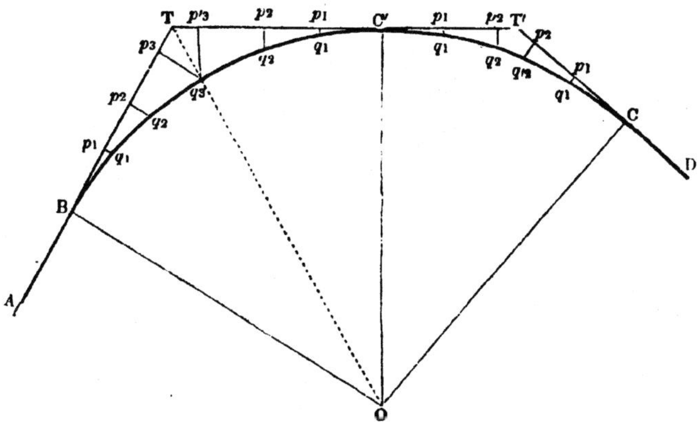

into two or more parts, and introducing one or more additional tangents. In the annexed figure the curve BC is divided into two parts in C′, at which point the tangent TC′T′ is drawn, meeting the tangents BT, T′C in T and T′, the tangents BT, TC′ being each taken =

$Bp + \frac{1}{8}p_3q_3$,* wherein Bp_3 is taken the nearest whole number of chains to $\frac{1}{8}$BO, and p_3q_3 is the offset corresponding to Bp_3. By thus taking the length of BT, TC′, the last offset on BT and the first one on TC will meet at q_3, the middle point of BC′, and therefore the distances of the ends of the offsets that form the curve will all be sufficiently near to equality, *i.e.*, to one chain. Also by trigonometry, BT : BO : : rad. : tan. ∠ BTO $=\frac{1}{2}$ ∠ BTC′; whence the ∠ BTC′ becomes known, and consequently the direction of the new tangent TC′T′ is also known. Having, therefore, determined BT and ∠ BTC′, the offsets may be laid out to p_3q_3; and having made $Tp_3 = Tp'_3$, the same offsets may be laid out in an inverted order on TC′; the order of the offsets being a second time inverted on C′T′, and a third time on T′C. If the distance of the middle offsets q_3q' be less or greater than 1 chain, that distance, and consequently the following ones, must be made equal to 1 chain, as in the preceding case.

Ex. Let BO = 240 ch., then $Bp_3 = \frac{1}{8}BO = 30$ ch., $p_3q_3 = \frac{Bp^{2}_{3}}{2BO} = \frac{30^2}{480} = 1{\cdot}875$ ch. : whence BT = TC′ = $Bp_3 + \frac{1}{8}p_3q_3 = 30 + \frac{1}{8}1{\cdot}875 = 30{\cdot}2344$ ch. $= 30{\cdot}23\frac{1}{2}$ ch. nearly : and by trig.

As 30·2344 (BT) log.	. . .	1·48050
: 240 (BO)	. . . . .	2·38021
: : rad.	. . . . .	10·00000
: tan. $\frac{1}{2}$ ∠ BTC′ = 82° 49′	.	10·89971
2		

∠ BTC = 165° 38′, whence the position of the tangent TC′T′ becomes known; it will be unnecessary in this case to find ∠ TT′C.

By table 4, the first offset on BT, viz. $p_1q_1 = 1{\cdot}65$ in., from which the other offsets may be found by multiplying by 4, 9, 16, &c., which may be laid out as in the last case, p_3q_3 being the 30th offset from B, and meeting the corresponding offset p'_3q_3 at q_3, the middle of the curve BC′. The method of making the distances equal on the second part C′C of curve, if required, is the same as in Case I.

NOTE.—If the curve be a long one, it may be divided, in this manner, into 3, 4, or 5 parts, according to its length.

---

* *Demonstration.*—By the similar triangles OBT, q_3p_3T, OB : $Bp_3 + p_3$T (BT) : : p_3q_3 : p_3T; but since $Bp_3 = \frac{1}{n}$ OB, or the nearest whole number to it, and since p_3T must always be very small compared with Bp_3, it may be rejected; ∴ OB : $\frac{1}{n}$OB (= very nearly to BT) : : p_3q_3 : $p_3T = \frac{1}{n}p_3q_3$, wherein n may be taken any whole number, 8 being the most convenient, in the exampleas.

Case III.

When the length of the curve exceeds the limit adopted in Case I., *but is considered too short to be divided into two parts, as in* Case II.

Let the curve be one of 80 chains radius, and between 32 and 38 chains in length, then to avoid the trouble of adding another tangent, the offsets beyond the 8th must be calculated from the following formulæ :—

$$p_7q_7 = r - \sqrt{r^2 - 9^2}$$
$$p_8q_8 = r - \sqrt{r^2 - 10}$$
$$\&c. = \quad \&c.$$

If the radius of the curve be 160 chains, and its length about 50 or 60 chains, the offsets must be calculated, as above, beyond the 15th. The formula $\frac{1^2}{2r}, \frac{2^2}{2r}, \frac{3^2}{2r}$, &c., is a very near approximation to the true lengths of the offsets within the limits assigned, but beyond these limits the errors of the offsets begin gradually to augment, till they become too considerable to be overlooked. See Demonstration to Case I.

PROBLEM IV.

To lay out the curve, by two theodolites, where water, swamps, quarries, or other obstructions prevent the use of the chain.

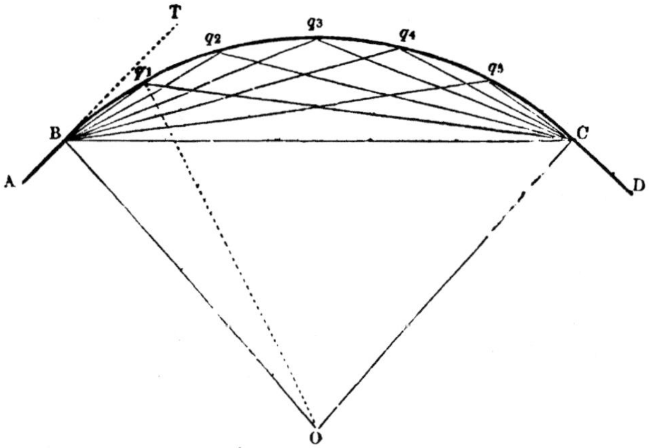

The position of the tangents AB, DC to the curve at its extremities B and C, its radius OB or OC, and the ∠ BOC = supplement of the angle made by the tangents when prolonged, being determined ; join

BC, and take several equidistant points q_1, q_2, q_3, &c., in the curve, from which points draw lines to B and C.

Take the $\angle\, q_1$CB = an arc whose sine is $\frac{\delta}{2r}$, wherein δ = Bq_1 = q_1q_2 = &c., and r = OB, or, because δ is, in practice, usually required to be = 1 chain, take $\angle\, q_1$CB = arc whose sine is $\frac{1}{2r}$. The several angles may then be arranged as below :—

$\angle\, q_1$CB = arc to sin. $\frac{1}{2r}$, and	$\angle\, q_1$BC = $\frac{1}{2}$ $\angle$ BOC − $\angle\, q_1$CB,
$\angle\, q_2$CB = 2 $\angle\, q_1$CB,	$\angle\, q_2$BC = $\frac{1}{2}$ $\angle$ BOC − 2 $\angle\, q_1$CB,
$\angle\, q_3$CB = 3 $\angle\, q_1$CB,	$\angle\, q_3$BC = $\frac{1}{2}$ $\angle$ BOC − 3 $\angle\, q_1$CB,
$\angle\, q_4$CB = 4 $\angle\, q_1$CB,	$\angle\, q_4$BC = $\frac{1}{2}$ $\angle$ BOC − 4 $\angle\, q_1$CB,
&c. = &c.	&c. = &c.

Therefore if theodolites be fixed at B and C, and the angles q_1CB, q_1BC be taken at the same time, the intersection of Bq_1 and Cq_1 will give the point q_1. In the same manner by taking the angles q_2CB, q_2BC, the intersection of Bq_2 and Cq_2 will give the point q_2, &c.

Ex. Let the radius OB = 80 chains, and the angle between the tangents AB, DC, when prolonged to meet, be 160°; then its supplement = $\angle$ BOC = 80°, and $\frac{1}{2}$ $\angle$ BOC = 40°. Also $\angle\, q_1$CB = arc to sine $\frac{1}{2r}\left(=\frac{1}{160}=\cdot 00625\right)$ = 0° 21′ 29″, whence $\angle\, q_1$BC = $\frac{1}{2}$ $\angle$ BOC − $\angle\, q_1$CB = 40° − 0° 21′ 29″ = 39° 38′ 31″; $\angle\, q_2$CB = 2 $\angle\, q_1$CB = 2 × 0° 21′ 29″ = 0° 42′ 58″ and $\angle\, q_2$BC = $\frac{1}{2}$ $\angle$ BOC − 2 $\angle\, q_1$CB = 40° − 0° 42′ 58″ = 39° 17′ 2″, &c. The angles would be best arranged for use as follows, the opposite ones to be taken at the same time.

$\angle\, q_1$CB = arc to sin. ·00625	= 0° 21′ 29″,	$\angle\, q_1$BC = $\frac{1}{2}$ $\angle$ BOC − $\angle\, q_1$CB	= 39° 38′ 31″;
$\angle\, q_2$CB = 2 $\angle\, q_1$CB	= 0° 42′ 58″,	$\angle\, q_2$BC = $\frac{1}{2}$ $\angle$ BOC − 2 $\angle\, q_1$CB	= 39° 17′ 2″;
$\angle\, q_3$CB = 3 $\angle\, q_1$CB	= 1° 4′ 27″,	$\angle\, q_3$BC = $\frac{1}{2}$ $\angle$ BOC − 3 $\angle\, q_1$CB	= 38° 55′ 33″;
$\angle\, q_4$CB = 4 $\angle\, q_1$CB	= 1° 25′ 56″,	$\angle\, q_4$BC = $\frac{1}{2}$ $\angle$ BOC − 4 $\angle\, q_1$CB	= 38° 34′ 4″;
&c. = &c.	= &c.	= &c.	= &c.

This list of angles must be continued till n $\angle\, q_1$CB can no longer be taken from $\frac{1}{2}$ $\angle$ BOC, and one angle in each column being taken at the same time, as $\angle\, q_1$CB at C and $\angle\, q_1$BC at B, will give the point q_1 : and so on for the other points q_2, q_3, &c. See the following notes.

NOTE 1. Where the obstructions are such as to prevent the stumps being put down at the consecutive points q_1, q_2, q_3, &c., it would be best to take every fourth angle, thus obtaining the points q_4, q_8, q_{12}, &c., which will be 4 chains apart on the curve, the intermediate points being left to be put in when the work of the line has progressed so far as to present a better opportunity. Or, if it should be thought preferable to avoid taking a multiplicity of angles, every fourth angle may be taken in any case, as this method is equally available whether obstructions exist or not.

Demonst. to Prob. IV.—The angles Bq_1C, Bq_2C, Bq_3C, &c., being in the same segment, are well known to be equal and constant, and also equal to half the supplement of $\angle BOC$, and consequently the sums of the angles of the triangles Bq_1C, Bq_2C, &c., adjacent to BC are equal to $\frac{1}{2}\angle BOC$, whence $\angle q_1BC=\frac{1}{2}\angle BOC-\angle q_1CB$, $\angle q_2BC=\frac{1}{2}\angle BOC-\angle q_2CB=$(because equal angles stand on equal arcs) $\frac{1}{2}\angle BOC-2\angle q_1CB$, &c. Also join Oq_1 and put $BO=r$, and $Bq_1=\delta$, then it is well known that $\frac{1}{2}\angle BOq_1=\angle q_1CB$, and because the sides BO, q_1O of the triangle BOq_1 are equal, $\sin.\frac{1}{2}\angle BOq_1=\sin.\angle q_1CB=\frac{\text{rad.}\times\delta}{qr}$; or, by taking rad.$=1$, and $\delta=1$ chain, as required in practice, $\sin.\angle q_1CB=\frac{1}{2r}$ or $\angle q_1CB=$arc to sine $\frac{1}{2r}$. *Q. E. D.*

PROBLEM V.

To lay out a railway curve by means of ordinates or offsets from its chord or chords, no material obstructions being supposed to exist on the concave side of the curve to prevent the use of the chain.

Let BC′C be a portion, or the whole, of a curve of a railway; AT, T′T′ and CD tangents to the curve at B, C′, and C; O the centre, OB the radius, BC′ and C′C chords of the curve; which chords must not exceed 40 chains, if the radius be 80 or 120 chains, but they may be 60

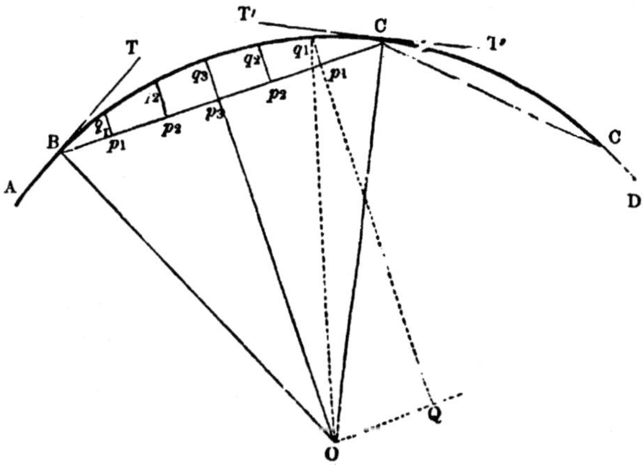

chains, if the radius exceed 120 (this limitation is necessary to prevent the offsets p_1q_1, p_2q_2, &c., being too long, as was observed with respect to the offsets from the tangents, in Prob. III.); and let the radius q_3o bisect the curve and chord in q_3 and p_3 respectively. Then from the right-angled triangle Bp_3O, in which BO and $Ap_3=$

$\frac{1}{2}$BC′ are given, the $\angle$ BOp_3 = $\angle$ TBC may be found, which determines the position of the chord BC′; also Op_3 may be found from the same triangle being = $\sqrt{OB^2 - BP_3^2} = \sqrt{OB^2 - \frac{1}{4}BC'^2}$. Put OB = r, Op_3 = s, and the $\frac{1}{2}$ chord Bp_3 = n chains, then

$$q_1p_1 = \sqrt{r^2-(n-1)^2} - s,^*$$
$$q_2p_2 = \sqrt{r^2-(n-2)^2} - s,$$
$$\&c. = \&c.$$
$$q_3p_3 = \sqrt{r^2-(n-3)^2} - s = r - s.$$

q_3p_3 being supposed to be the offsets at the middle point of the curve BC′, not the third offset as shown in the figure, it being impossible to draw all the offsets without confusing the figure, or making it unnecessarily large. After reaching the middle point q_3 of the curve, the same offsets are repeated in an inverted order till the curve shall have been set out to C′. The same operation may be repeated as often as necessary, till the whole curve be completed, observing to make the $\angle$ BC′C = 2 complement of $\angle$ TBC′, which has been already found. See the following notes :—

NOTE 1. When BC′ is the whole curve, and its chord Bp_3C includes a fractional part of a chain, the distance of the offsets on each side of the middle of the curve will be less than one chain ; therefore that distance, and consequently the following ones, must be made equal, as was shown with respect to the distances in Problem III.

2. If the last chord, which suppose to be C′C, be less than the preceding chord or chords, the $\angle$ OC′C, must be found, and added to the $\angle$ OBC′ r OC′B, which will give the $\angle$ BC′C, showing the direction of the chord C′C.

3. This method of laying out the curve is seldom used, on account of the calculations it involves. It may, however, be used with advantage where a winding river or cliff is close to the convex side of the curve, or protrudes in some places a little through the curve ; thus preventing the use of any other method, except that in Problem IV., which requires two theodolites.

Definition of the compound curve. (See figures to Prob. VI.)

The compound curve BCC′C″, joining the tangents AB, DC″, is composed of three circular arcs BC, CC′, C′C″, having common normals, OC, O′C′ at their points of junction C, C′; and therefore common tangents at the same points, the radii of the three portions of the curve being respectively OB = OC, O′C = O′C′, and O″C′ = O″C″. This kind of curve is adopted where the line is required to pass through

* *Demonstration.*—Draw OQ parallel to BC′, q_1Q ⊥ to OQ, and join q_1O ; and let n = number of chains in $\frac{1}{2}$BC′ or p_3C′, and C′p_1 = 1 chain. Then p_3p_1 = OQ = $n-1$; and (Euc. I. 47) $q_1Q = \sqrt{r^2-(n-1)^2}$; whence $q_1p_1 = \sqrt{r^2-(n^2-1)^2} - p_1Q$: but $p_1Q = p_3O = s$, $\therefore$ $q_1p_1 = \sqrt{r^2-(n-1)^2} - s$. Similarly, q_2p_2 is found = $\sqrt{r^2-(n^2-2)^2} - s$, &c. *Q. E. D.*

given points, as C and C′, to avoid obstructions, or where a principal station or *terminus* is at or near C″; in the latter case the radius O″C″ may, if required, be less than 80 chains.

The compound curve may consist of two, three, or more portions of different arcs; thus the curve *bcc′* consists of two portions, *bc*, *cc′*.

PROBLEM VI.

1. *To find the several radii of the compound curve mechanically.*

Let AB, DC″ be the tangential portions of the line, which are required to be joined by a curve passing through the points C, C′, *the point* C″ *not being given.* Select a curve-ruler such that, being applied to touch AB at B, it may also pass through C; if this curve do not pass through C′, but through some other point E, another curve ruler of less radius, in this case, must be selected, and such that it may touch the arc BC at C without cutting it or its prolongation towards E, and also pass through the point C: if this curve cut the prolongation C″T′ of the tangent DC″, another curve-ruler of less radius than the last one must be selected, and such that it may touch curve CC′ at C′ and the tangent DC″ at C″: thus completing the curve BCC′C″. The radii OC, O′C′, O″C″ may be determined from the curve-rulers, as in Case I. Prob. I.

2. *To find the radius c′o′ of the compound curve bcc′ geometrically, the starting-point b and the radius bo being given.*

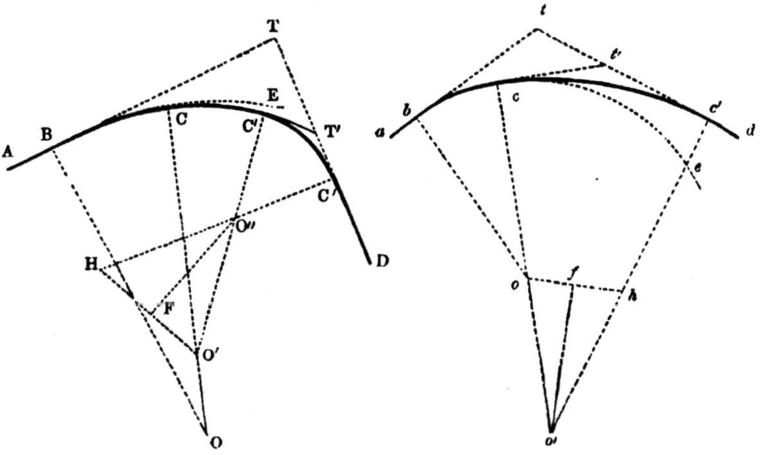

From the given point *b* in the tangent *ab* draw the given radius *bo* ⊥ to *ab;* and draw the curve to some point *c*, where it is found convenient to change the radius: draw the radius oc, and $\perp$ thereto draw ct', meeting the tangent dt in t'; make $t'c' = t'c$, and from c' draw $c'o' \perp$ to tc' meeting co, prolonged if necessary in o'; then o' is the centre of the arc cc' of the curve, conformable to the nature of tangents.

The method of constructing the curve, when it consists of three or more parts, is sufficiently obvious.

3. *One of the two radii of the compound curve and its starting and closing points being given, to find the other radius.*

Let ab, $c'd$, be the tangents, b and c' the starting and closing points of the curve. Draw the perpendiculars $bo = c'h$ = given radins to the tangents; join oh, and bisect it in f; draw $fo \perp$ to oh, meeting $c'h$ prolonged in o'; join $o'o$, and prolong it till $oc = c'h$: then the points o, o' are the centres of the arcs bc, cc', which constitute the compound curve, $o'c = o'c'$ being the radius required.*

NOTE.—In the compound curve BCC'C'', where the radius C''O'', which is to be found, is less than the preceding radius CO, the $\perp$ C''H is made = CO; HO' is joined and bisected in F; and FO'' drawn $\perp$ to HO', meeting C''H in O'', which is the centre of the arc C'C'', &c.

Definition of the serpentine or S *curve.* (See figure to Prob. VII.)

The serpentine curve BGC is used in railways, when obstructions or some other cause render its adoption preferable; it consists of two circular arcs, having their convex sides turned in opposite directions, like the letter S, whence it is sometimes called the S curve; the two portions BG, GC of the curve have a common normal OGO' at their point of junction G, and therefore a common tangent at the same point. This curve affords the most easy means of joining two parallel, or nearly parallel, portions of a line of railway.

PROBLEM VII.

1. *When one radius and its tangential point are given, to find the other radius and tangential point of the serpentine curve geometrically.*

From the given tangential point C draw the given radius CO $\perp$ to the tangent CD, and draw the curve CG to some point G where it is found convenient that it should have its point of contrary flexure;

* *Demonstration.*—Since $of = fh$, and fo' is $\perp$ to oh, $oo' = o'h$; also bo was made $= c'h = oc$; $\therefore$ $o'c = o'c'$ and the normal $o'oc$ is common to the arcs of the curve. *Q. E. D.*

through OG draw the normal OGO′; from G draw GT ⊥ to OGO′, to meet the tangent AT; make TB = TG; and draw BO′ to AT, meeting

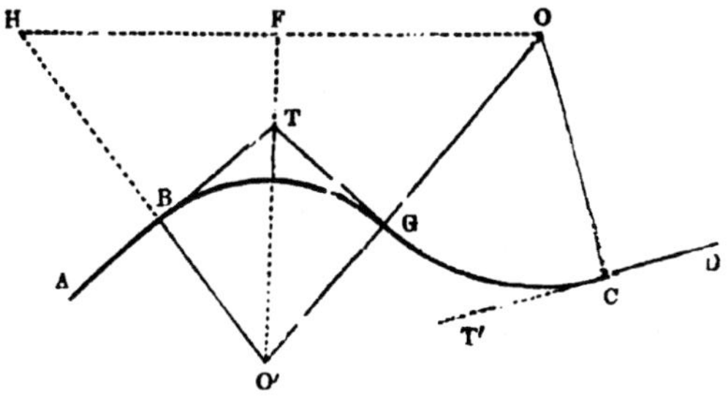

OGO′ in O′; then O is the centre, and O′B = OG is the radius of the curve BG, as is evident from the nature of tangents.

NOTE.—The radius O′B and tangential point B may be found mechanically, *i.e.*, by the curve-rulers, as in Problems I. and VI.

2. *When the tangential points and one of the radii of the serpentine curve are given, to find the other radius geometrically.*

From the given tangential points C and B draw CO, BH, respectively ⊥ to the tangents CD and BA, and equal to the given radius; join OH, and bisect it in F; draw FO′ ⊥ to OH, meeting HB prolonged in O′, and join O, O′; making O′G = O′B; then O′ is the centre, and O′B = O′G is the radius of the portion BG of the curve, as required.*

NOTE 1. The radius BO′ may be found mechanically as in the preceding case.

2. The radius O′B may be found from the following formula, wherein $\delta = BC$, $r =$ given radius OC, $\alpha = \angle TBC$, and $\alpha' = \angle T'CB$.

$$O'B = \frac{\delta\,(\delta - 2r\,\text{sin.}\,\alpha')}{2\left(\delta\,\text{sin.}\,\alpha + 2r\,\text{sin.}^2\dfrac{\alpha-\alpha'}{2}\right)}.$$ See investigation and fig. to Problem VIII.

Ex. Let $\delta = 200$ chains, $r = 110$ chains, $\alpha = 45°$, and $\alpha' = 20°$; the sin. $\alpha =$ ·70711, sin. $\alpha' =$ ·34202, $\dfrac{\alpha - \alpha'}{2} = \dfrac{45° - 20°}{2} = 12°\,30'$, the sine of which is ·21644; and by the formula $O'B = \dfrac{200\,(200 - 220 \times \cdot 34202)}{2\,(200 \times \cdot 70711 \times 220 \times \cdot 21644^2,} -$

$\dfrac{200 \times 124\cdot 7556}{2 \times 151\cdot 7282} = 82\cdot 22$ chains.

It may thus be readily ascertained whether the required radius O′B is greater, equal to, or less than 80 chains; if less, the given radius r ought either to be diminished, or the distance BC of the tangential points ought to be increased, according to circumstances, in order that O′B may be of the required length, assuming that the portion BG of the curve is not near a *terminus* or principal station.

---

* The *demonstration* in this case is similar to the one given to Problem VI.

PROBLEM VIII.

1. *When the two portions of the serpentine curve have the same radius, to determine that radius geometrically, the tangential points and their distance being given.*

Let AB, DC be the tangents, B and C the tangential points, and BC the given distance. Draw $Bo = Co'$ respectively ⊥ to AB, DC, and of any convenient length; through o, parallel to BC, draw oq

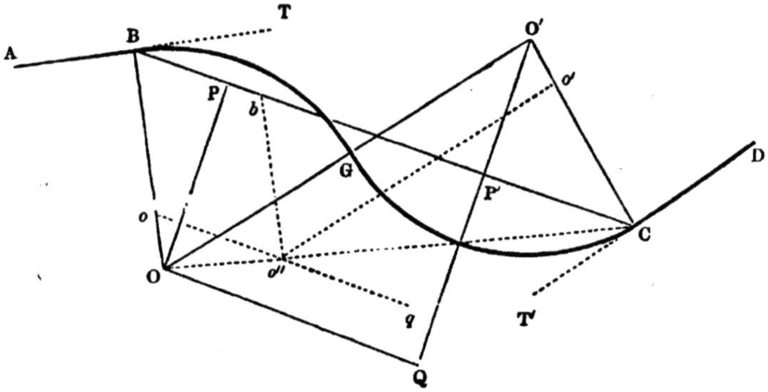

indefinitely; with the compasses apply $o'o'' = 2Co' = 2Bo$; through C, o'' draw $Co''O$, meeting BO prolonged in O; and through O, parallel to $o''o'$, draw OO, meeting Co' prolonged in O′; then O and O′ are the centres, and OB and O′C are the equal radii of the serpentine curve BGC, the common normal of the portions BG, GC of the curve being OGO′ = 2BO = 2CO′.*

2. *To find the common radius of the two portions of the serpentine curve by calculation, the same things being given as in the preceding case, and the angles* TBC, T′CB.

Put BC $= \delta$, BO $=$ CO′ $= r$, ∠ TBC $= \alpha$, and ∠ T′CB $= \alpha'$; then

$$r = \frac{\delta}{\text{sin. } \alpha + \text{sin. } \alpha' + 2 \text{ sin. of arc to cos. } \frac{1}{2} (\text{cos. } \alpha + \text{cos. } \alpha')}.\dagger$$

* *Demonstration.*—Draw $o''b$ parallel to OB; then by similar triangles,

$Co' : o'o'' = 2Co' : :$ OO′ : 2CO′ = OO′,

$Co'' :$ CO $: : Co' = Bo = bo'' :$ CO′

∴ Co'' : CO $: : Co' = Bo = bo''$: BO = CO′. *Q. E. D.*

† *Investigation.*—Draw PO, P′O′ ⊥ to BC; through O draw OQ parallel to BC, meeting O′P′ prolonged in Q: then OO′ $= 2r$, BP $= r$ sin. α, OP $=$ P′Q $= r$ cos. α, CP′ $= r$ sin. α', O′P′ $= r$ cos. α', O′Q $=$ O′P′ $+$ P′Q $= r$ (cos. α + cos. α'): whence, from the right-angled triangle OQO′, sin. ∠ O′OQ $\frac{1}{2}$ (cos. α + cos. α'), which is therefore

Ex. Let $\delta = 200$ chains, $\alpha = 27°$, $\alpha' = 50°$: then from a table of nat. sines sin. $\alpha = \cdot45399$, its cos. $= \cdot89101$; sin. $\alpha' = \cdot76604$, its cos. $= \cdot64279$: whence $\frac{1}{2}$ (cos. α + cos. α') $= \frac{1}{2}$ ($\cdot89101 + \cdot64279$) $= \cdot76690 =$ cos. and 2 sin. of arc to cos. $\cdot76690$ is $1\cdot28334$; therefore

$$r = \frac{200}{\cdot45399 + \cdot76604 + 1\cdot28334} = \frac{200}{2\cdot50337} = 79\cdot89 \text{ chains, or nearly}$$

80 chains, the radius required.

NOTE.—The method of forming the serpentine curve with a common radius is much to be preferred to any other, when the nature of the ground will admit of its being done; and more especially so, when the *data*, as in the preceding example, will only just give a common radius of 80 chains : for, if the radius of one of the portions of the curve had been taken greater than 80 chains, the other radius would have necessarily been less than 80 chains.

The Continental engineers, in laying out serpentine curves, usually place, if possible, a short *straight* piece of line between the two members of the curve, to ease the transfer of the train from one direction of curvature to the other.

PROBLEM IX.

To make a given deviation HQ *from a straight portion of a line of railway* AHD *by means of three curves,* BG, GQG′, G′C, *having their radii* OB, O′Q, O″C, *all equal, in order that the lateral works of the line may avoid the building, or other obstruction b, which is close to the centre of the straight portion of the line.*

1. *Construction.*—From the given point H draw HQ = given deviation ⊥ to AD ; on QH prolonged, take QO′ = OP = given radius ; with the compasses apply PB = QP = twice given radius ; draw

given, whence the comp. of ∠O′QO = ∠OO′Q is known, and may be thus expressed, sin. ∠OO′Q = sin. of arc to cos. $\frac{1}{2}$ (cos. α + cos. α'). Whence OQ = PP′ = OO′ × sin. ∠OO′Q = $2r$ × sin. of arc to cos. $\frac{1}{2}$ (cos. α + cos. α'), δ = BC = BP + PP′ + P′C = r sin. α + r sin. α' + $2r$ sin. of arc to cos. $\frac{1}{2}$ (cos. α + cos. α'), from which

$$r = \frac{\delta}{\sin.\ \alpha + \sin.\ \alpha' + 2\ \sin.\ \text{of arc to cos.}\ \frac{1}{2}\ (\cos\ \alpha + \cos.\ \alpha')}. \quad Q.\ E.\ I.$$

When the radii are unequal, and one of them, as r = O′C is given, and the other R = BO is required; then O′O = R + r, BP = R sin. α, PO = P′Q = R cos. α, O′Q = R cos. α + r cos. α', and OQ = PP′ = $\sqrt{O'O^2 - O'Q^2} = \sqrt{(R+r)^2 - (R \cos.\ \alpha + r \cos.\ \alpha')^2}$, δ = R sin. α + r cos. α' + $\sqrt{(R+r)^2 - (R \cos.\ \alpha + r \cos\ \alpha')^2}$. By transposing and squaring, and remembering that sin.² + cos.² = 1, &c., there results $\delta^2 - 2\delta$ (R sin. α + r sin. α') = $2Rr$ (1 − cos. α cos. α' − sin. α sin. α') = $2Rr$ (1 − cos. $\overline{\alpha - \alpha'}$) = $4Rr \sin.^2 \frac{\alpha - \alpha'}{2}$, whence

$$R = \frac{\delta\ (\delta - 2r \sin.\ \alpha'}{2\left(\delta + 2r \sin.^2 \frac{\alpha - \alpha'}{2}\right)}.$$ This formula is used for finding the value of BO′, Note 2, Problem VII., where the symbols are defined. *Q. E. I.*

BO ⊥ to AD; through O′ draw O′GO parallel to PB, meeting OB in O; make CH = HB; and join CQ, QB, the latter cutting OO′ in G, and the former cutting O′O″ (which is similarly drawn to OO′) in G′: then B and C are the starting and closing points of the curve, of which the separate portions are BG, GQG′, G′C, and the chords BG, GQ, QG′, G′C are all equal.*

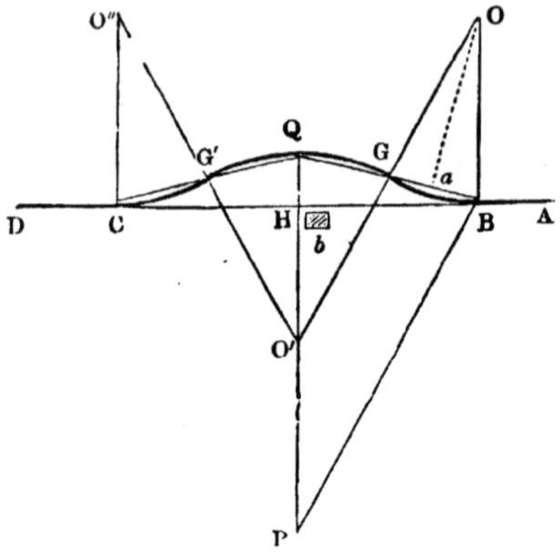

2. *Calculation.* — Take BH = HC = $\sqrt{QH\ (4BO - QH)}$, and BG = GQ = QG′ = G′C = $\sqrt{BO \,.\, QH}$, which chords of the arcs BG, GQ, &c., thus become known; and, since the common radius BO is given, the construction of the curve is obvious.†

Ex. Let the given deviation QH = 2 chains, and the common radius BO = 85 chains; then BH = HC = $\sqrt{2(340-2)} = \sqrt{676}$ = 26 chains, and BG = GQ = &c. = $\sqrt{2 \times 85} = \sqrt{170}$ = 13·04 chains.

REMARKS ON LAYING OUT THE CURVES IN THE FOUR LAST PROBLEMS.

Having in the four last problems given various methods of determining the radii and common normals, indicating the positions of the tangent points of the parts of the compound, serpentine, and deviation curves, the method of laying out the curves themselves by Problems II., III., IV., or V., according to circumstances, will be readily seen, recollecting that when junction-points of curves of different radii occur, as CC′, first fig. to Prob. VI., to commence the operation afresh, by using the radii and tangents of the respective portions of the curve.

* *Demonstration.*—Because QP, BO are parallel, as are also O′O, PB; O′P = BO = (by const.) QO′ = given radius, and O′O = PB = 2QO′ = 2GO′; ∴ GO′ = GO = QO = BO, which are similarly proved to be = G′O′ = G′O″ = CO″. *Q. E. D.*

† *Demonstration.*—Draw Oa perpendicular to BQ, bisecting it in a; then by similar triangles BO : Ba = $\frac{1}{4}$BQ : : BQ : QH, whence $BQ^2 = 4BO \,.\, QH$, or $BG^2 = \frac{1}{2}BQ^2$ = BO . QH, or BG = $\sqrt{BO \,.\, QH}$. Also $BH^2 = HC^2 = BQ^2 - QH^2 = 4BO \,.\, QH - QH^2 = QH\,(4BO - QH)$, or BH = HC = $\sqrt{QH\,(4BO - QH)}$. *Q. E. D.*

SECTION II.

RAILWAY EARTH WORKS.

On setting out the width of ground for a railway.

After the centre stumps of the railway have been put down, which, as before observed, are usually at the distance of one chain apart, the line must next be carefully levelled, and the number of the stumps entered in the Level-Book, in a vertical column; and opposite each number, in a second column, the depth of the cuttings or embankments (see Level-Book, page 441); and in a third column, the computed or horizontal half-width of the surface cuttings as found by Problems I. and II. following; the depths of the cuttings and embankments being estimated from the balance line, which is 2 feet below the line of the rails or gradients, the 2 feet being filled up with gravel to form the way, and the beds for the sleepers of the rails.

The side stumps are next to be put down, which must be placed two on each side of every centre stump, in a direction perpendicular to the length of the line; or, if the line be curved, in a direction perpendicular to the tangent to the curve at the centre stump: the two interior stumps, *i.e.*, those next to the centre one, to mark the width of the cuttings, and the two exterior ones to mark the ditches of the side fences. The distances of every two of the interior stumps are to be entered in the Level-Book, opposite the number of the centre stump, in the main section, in order to ascertain the quantity of cuttings for the contractor; and the distances of every two of exterior stumps from the centre one are to be similarly entered, to ascertain the quantity of land which will be required from each proprietor for the works of the line; which last may be calculated accurately by the method of equidistant ordinates. (See Problem VI.)

PROBLEM I.

To set out the width when the surface of the ground is laterally on the same level as the intended railway.

From the centre stump, perpendicular to the direction of the line, set out half the bottom-width for the cutting, to which add the width of the side-fence, putting down a stump at each distance; then repeat the operation on the other side of the line.

Ex. If the bottom-width of the railway be 30 feet, and the breadth of one of the side fences be 12 feet, required the widths for cutting and for fences.

$30 \div 2 = 15$ ft. = dist. of side-stump from centre for cutting.

$15 + 12 = 27$ ft. = dist. of side-stump from centre for fence.

Therefore, $15 \times 2 = 30$ = whole width for cutting.

and $27 \times 2 = 54$ = whole width for fences.

PROBLEM II.

To set out the width of cuttings or embankments, when the surface of the ground is laterally level, and at a given height above the level of the intended railway, the ratio of the slopes being given.

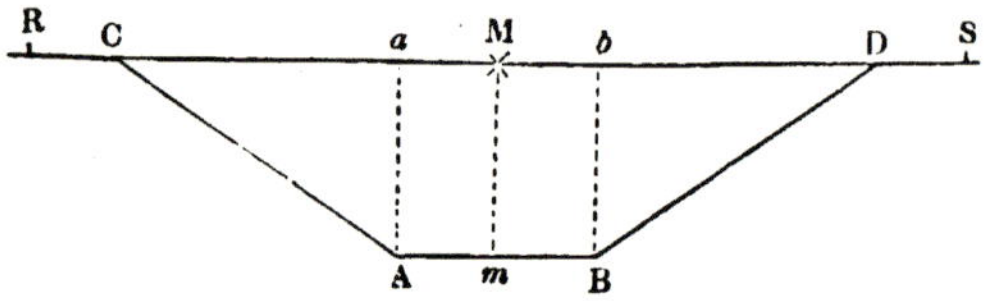

Let ABCD be a cross-section of the cuttings, RS the horizontal surface of the ground, AB the bottom width, AC, BD the slopes, $Mm = Aa = Bb$ the perpendicular depth, and M the middle stump. Multiply the given depth Mm by the ratio of the slopes, to which add half the bottom-width Am, or aM: set out this distance from M to C for the half-width of the cuttings, to which add the width of the side fence for the whole half-width, repeating the same operation on the other side of M.

Ex. If the bottom-width AB or ab be 30 feet, the depth of the cuttings 28 feet, and the ratio of the slopes $1\frac{1}{2} : 1$, and the width of one of the side fences 12 feet, required the width of the cuttings, and of the land for the works of the railway.

$(28 \times 1\frac{1}{2}) + \frac{30}{2} = 42 + 15 = 57$ feet = MC = MD = $\frac{1}{2}$ width of cuttings; and $57 + 12 = 69$ feet = MR = MS = $\frac{1}{2}$ width of land. The doubles of which are the whole widths.

If w = bottom-width = AB, a = depth of cuttings = Mm, f = width of one of the side-fences = RC, and the ratio of the slopes $r : 1$. Then

$$1 : r :: a : aC = ar, \text{ hence } ar + \tfrac{1}{2}w = MC = MD,$$

$$ar + \tfrac{1}{2}w + f = MR = MS,$$

$$\text{and } 2\,(ar + \tfrac{1}{2}w + f) = w + 2ar + 2f = RS.$$

Construction.—Draw AB = given width = 30 feet, perpendicular to which, at its middle point m, draw mM = given depth = 28 feet; through M parallel to AB draw RMS; through A and B draw Aa, Bb parallel to Mm; make $aC = bD = 1\frac{1}{2}$ times Mm, join AC, BD, and make

RC, DS = width of one of the side-fences = 12 feet. Then ABDC is a cross section of the cuttings, CD the surface-width thereof, and RS the whole surface required for the railway.

By inverting the last figure, so that ABDC may represent the cross-section of an embankment, it will be readily seen that the same method will apply, for setting out its half-widths, MC, MR, as that just given for the cuttings; as the dimensions are the same in both cases.

NOTE 1. The ratio of the slopes is the proportion of the horizontal line ca to the depth Aa or Mm. Thus, when the ratio is 1 : 1, ca = Aa : when the ratio is 2 : 1, ca = 2Aa, &c. This ratio depends on the nature of the material through which the cuttings are made; if it is close-jointed hard rock, the ratio is $\frac{1}{4}$ or $\frac{1}{2}$ to 1; if soft, or loose-jointed rock, or strong clay, the ratio is 1 or $1\frac{1}{2}$ to 1; if springy ground, or loose sand, the ratio is 2 or $2\frac{1}{2}$ to 1.

2. The computed half-widths, in the third column of the Level-Book (following Problem V.), are found by this Problem.

PROBLEM III.

To set out the width of the cuttings, when the surface of the ground is laterally sloping, the height of the centre stump above the level of the intended railway, the ratio of the slopes, and the lateral fall or rise of the ground in a given horizontal distance being given.

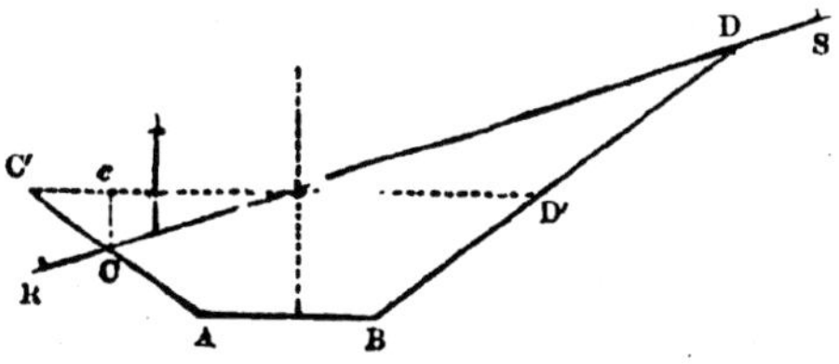

Let ABDC be a cross-section of the cuttings, RS the sloping surface of the ground, AB the bottom-width, Mm the depth of the cuttings, M the middle stump, AC, BD the slopes, and C′D′ a horizontal line passing through M, MC′ = MD′ being the computed half-widths.

Having set the level so that by turning the telescope two or three chains' length of the line may be seen, if possible, on both sides of it, place a levelling-staff at M, and another at q, and observe the level readings on them the difference of which will be pq; measure with the tape-line in feet the distance Mq on the slope and the horizontal distance Mp; take the computed half-width from the Level-Book, or find it by Problem II., and multiply it by the distance Mq on the slope, and reserve the product. Add and subtract the product of the difference of the staff-readings, and the ratio of the slopes, to and from the horizontal distance Mp, and reserve the sum and difference; divide the reserved product by the reserved sum for the corrected half-width MC, and by the reserved difference for the corrected half-width MD.

Ex. Let the depth of the cuttings at the centre-stump M be 20 feet, the bottom width AB = 30 feet, the distance Mq on the slope = 25 feet, the distance Mp on the level = 24 feet, the difference of the stave readings $pq = 7$ feet, and the ratio of the slopes $1\frac{1}{2} : 1$; required the corrected half-widths MC, MD.

$$20 \times 1\tfrac{1}{2} + \frac{30}{2} = 45 \text{ feet} = \text{computed half-width}$$

$$\begin{array}{r} 25 \\ \hline 1125 \text{ reserved product.} \end{array}$$

$$\begin{array}{l} \quad\quad\quad\quad 24 \\ 7 \times 1\tfrac{1}{2} = 10\tfrac{1}{2} \end{array}$$

reserved sum $34\frac{1}{2})1125(32{\cdot}6$ feet = cor. ½ width MC

reserved diff. $13\frac{1}{2})1125(83{\cdot}33$ feet = cor. ½ width MD.

Put MC′ $= b$, $pq = h$, M$q = s$, M$p = l$, MC $= x$, MD $= x'$, and the ratio of the slopes $r : 1$.

$$\text{Then } x = \frac{bs}{l + rh} = \text{MC*}$$

$$\text{and } x' = \frac{bs}{l - rh} = \text{MD}.$$

If the numbers in the preceding example be substituted in these formulæ, the work will stand thus :

$$\text{MC} = \frac{bs}{l + rh} = \frac{45 \quad 25}{24 + (7 \quad 1\frac{1}{2})} = \frac{1125}{34\frac{1}{2}} = \frac{2250}{69} = 32{\cdot}6 \text{ feet.}$$

$$\text{MD} = \frac{bs}{b - rh} = \frac{45 \quad 25}{24 - (7 \quad 1\frac{1}{2})} = \frac{1125}{13\frac{1}{2}} = \frac{2250}{27} = 83{\cdot}33 \text{ feet;}$$

to each of which values of the corrected half-width, the width of

* *Demonstration.*—Draw Oc ⊥ to C′D′; then by similar triangles, $s : h :: x : \frac{hx}{s} =$ cc, hence $c'c = \frac{rhx}{s}$, and M$c = c'$M $- c'c = b - \frac{rhx}{s}$; again, by similar triangles $l : h :: b - \frac{rhx}{s} :$ cc $= h$, $\frac{(bs - rhs)}{ls} = \frac{hx}{s}$, whence $x = \frac{bs}{l + rh} =$ MC; and in a similar manner is found $x' = \frac{bs}{l - rh} =$ MD. *Q. E. D.*

Cor. When the difference of the level readings is so large that the horizontal distance l cannot be conveniently measured, the value of $l = \sqrt{s^2 - h^2}$ must be substituted, thus giving

$$\text{MC} = \frac{bs}{\sqrt{s^2 - h^2} + rh}, \text{ and MD} = \frac{bs}{\sqrt{s^2 - h^2} - rh}.$$

one of the side-fences must be added for the whole breadth of land required.

NOTE 1. When the difference of level-readings is small, l may be taken $= s$ without material error.

2. A little practice will make this method of finding the corrected half-widths very easy. It is evidently much superior to the method of finding the same things by the blundering approximations often in use, especially where the difference of the levels of C and D is very considerable, and the surface of the ground has a regular slope, which often happens to be the case, or very nearly so; or, if the surface be a little uneven, a skilful engineer, by applying the tape-line in the general direction of the surface of the ground, will be readily able to correct the difference pq of the level-readings, so as to adapt it to his purpose. If, however, the surface of the ground be very uneven, which sometimes happens to be the case, it will be best to use the method of approximation, as in Problem V.

Construction.—Draw the cross-section ABD′C′ as in the preceding Problem; make Mp = given horizontal distance; draw pq perpendicular to C′D′ and equal to the difference of the level readings; and through Mq draw RS, cutting AC′ in C and BD′ prolonged in D: then ABDC is the required cross-section of the cuttings, RC, DS being the breadths of the side fences.

By reversing the cross-section, it will be seen that the preceding calculation and construction will also apply to an embankment; but in this case the lesser distance MC must be measured up the slope, and the greater MD down it.

PROBLEM IV.

To set out the widths when the surface of the ground is laterally sloping, and when the cross-section of the works consists partly of a cutting and partly of an embankment, the things given being the same as in the preceding Problem.

Let ACPBD be a cross-section of the works of a railway, consisting partly of the cutting BPD, and partly of the embankment APC; AB the bottom-width; M the centre stump, and Mm the depth of the cutting; RS the sloping surface of the ground, &c.

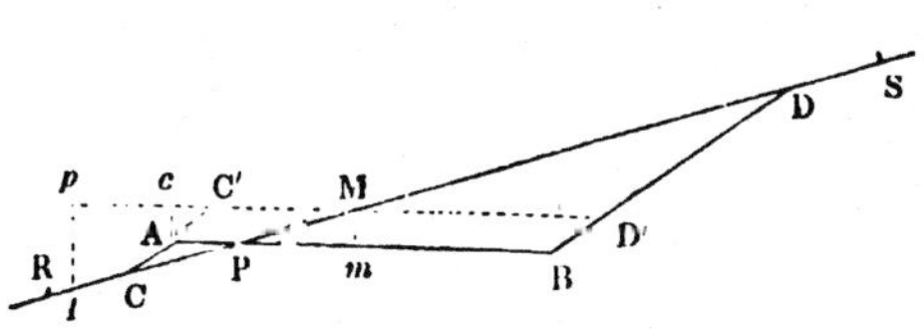

When the embanked part, APC, is less than half the bottom-width Am, proceed to find, as in Problems II. and III., the computed half-width MD′, and from thence the corrected half-width MD. But to find the corrected half-width MC (which consists partly of an embankment), multiply the distance MD (just found) by the difference of the bottom-width and the estimated half-width, and divide the product by the estimated half-width, and the quotient is the corrected half-width MC.*

Ex. Let the bottom-width AB $= 30$ ft., the depth M$m = 3$ ft., the ratio of the slopes $1\frac{1}{2} : 1$, and the difference of level-reading 7 ft., at the distances of 25 and 24 ft. from the centre-stump on the surface-slope, and on the level respectively; required the corrected half-widths MD and MC.

By Problem II. $3 \times 1\frac{1}{2} + \dfrac{30}{2} = 19\frac{1}{2}$ ft. = estimated half-width MD′,

By Problem III. $\dfrac{19\frac{1}{2} \times 25}{24 - (7 \times 1\frac{1}{2})} = \dfrac{975}{27} = 36{\cdot}11$ ft. = corrected half-width MD.

By Problem IV. $\dfrac{36{\cdot}11 \times (30 - 19\frac{1}{2})}{19\frac{1}{2}} = 19{\cdot}44$ ft. = corrected half-width MC.

If the same symbols be used for the given parts in this example, as in Problem III., and w = bottom-width: then

$$\text{MD} = \frac{bs}{l - rh} = \frac{19\frac{1}{2} \times 25}{24 - (7 \times 1\frac{1}{2})} = 36{\cdot}11 \text{ ft.} = \text{corrected half-width.}$$

$$\text{MC} = \frac{(w - b)s}{l - rh} = \frac{(30 - 19\frac{1}{2})25}{24 - (7 \times 1\frac{1}{2})} = 19{\cdot}44 \text{ ft.} = \text{corrected half-width.}$$

Note.—This operation for finding MC, it will be seen, is different from the preceding one. See *Demonstration.*

Construction. The operation for this Problem is the same as that for Problem III., excepting that AC is drawn parallel to BD.

By reversing the cross-section, it will be readily seen that the same calculation and construction will apply in the case of APC being a cutting, and PBD an embankment, observing that the corrected half-width for the cutting is the shorter distance, and *vice versâ.*

* *Demonstration.* Draw AC ⊥ to C′D′ prolonged, and let pq, which is ⊥ to pC′D′, be the difference of level-readings at M and q; then, if a = Mm, MC $=$ A$m = \frac{1}{2}w =$ $\frac{1}{2}$ bottom width, the other symbols being the same as in the Demonstration to Problem III., $ar =$ CC′ and C′M $=$ MC $-$ CC′ $= \frac{1}{2}w - ar$; whence, by the Demonstration above referred to, MC $= \dfrac{s\,(\frac{1}{2}w - ar)}{l - rh}$: but $b = \frac{1}{2}w + ar$, or $ar = b - \frac{1}{2}w$; this value being substituted in that of MC, gives $\dfrac{s\,(w - b)}{l - rh} =$ MC. Also $\dfrac{s\,(w - b)}{l - rh} =$ MC $= \dfrac{bs}{l - rh} \times \dfrac{w - b}{b}$, which is the rule given in words at length. *Q. E. D.*

PROBLEM V.

To find the width of the cuttings when the surface of the ground is laterally very uneven.

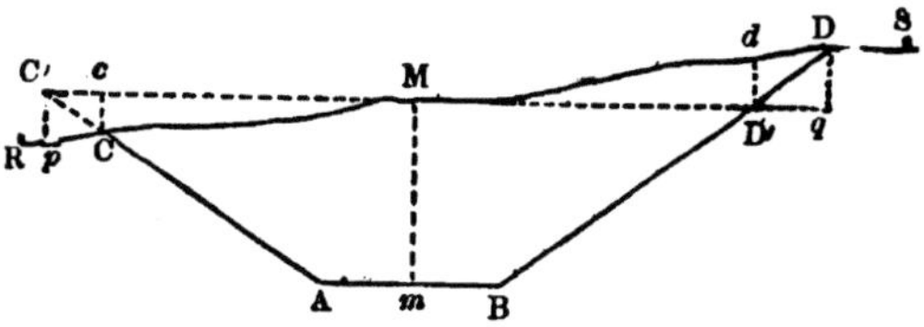

Let ABDMC be a cross-section of the cuttings, M*m* the depth, C′M = MD′ = estimated half-width, CDM the uneven surface of the ground, &c. The solution of this Problem will be best effected by giving an example in numbers.

Let AB = 30 ft., M*m* = 30 ft., and the ratio of the side-slopes as $1\frac{1}{2} : 1$; then $MC' = MD' = (30 \times 1\frac{1}{2}) + (\frac{1}{2} \times 30) = 60$ ft. Measure from M horizontally the distance M*d* = 60 ft; the point *d* is directly above D′. Place levelling-staves at M and *d*, and observe the difference of the readings at M and *d*, which, in this base, is 7 ft.; whence $7 \times 1\frac{1}{2} = 10\frac{1}{2}$ ft. = approximate distance *d*D, and M*d* + *d*D = D = 60 + $10\frac{1}{2}$ = $70\frac{1}{2}$ ft. Now place a levelling-staff D, and observe the reading, which is found to be 7·8 ft. greater than that at M, or 0·8 ft. greater than that at *d*; whence 0·8 $1\frac{1}{2}$ = 1·2 ft., and consequently $70\frac{1}{2} + 1{\cdot}2 = 71{\cdot}7$ ft., which is a still nearer approximation to the true distance MD or M*q*, it being measured horizontally. The operation for finding M*c*, or MC measured horizontally, is the same as the preceding, excepting that the product of the staff-readings is subtracted from the estimated half-width, &c. In this manner the distance M*c* = horizontal distance MC is found to be 52·6 feet.

NOTE 1. The widths RC, DS, of the side-fences, must be added to the above results for the whole width.

2. If the difference of the staff-readings at M and *d* be very large, it will require three or four approximations similar to those given in the preceding example to find the true corrected half-width.

Construction. Take the levels of the several undulations of the surface CMD, making C′MD′ the *datum*-line, and draw the cross-section ABDMC by the methods already given.

By reversing the cross-section ABDMC, its application to an embankment is obvious, observing also to reverse the distances MC, MD, as previously noticed.

3. The truth of the method of approximation used in this example is too obvious to require a demonstration.

LEVEL-BOOK.

	No. of Stump.	Depth of Cuttings or Embank-ments.	Computed Half-width.	Corrected half-widths for edge of Cutting or foot of Embankment.		Whole widths including Fences, each Nine Feet in Width.
				North.	South.	
		Feet.	Feet.	Feet.	Feet.	Feet.
Cutting.	216	30·00	60·00	52·60	71·70	142·30
	217	3·00	19·50	19·44	36·11	73·55
	218	28·00	57·00	57·95	56·08	133·55
	219	19·68	44·42	44·42	44·42	106·82
	220	20·00	45·00	32·60	83·33	133·93
Embankt.	221	16·08	39·12	39·72	59·12	116·84
	222	30·00	60·00	54·24	68·05	140·29
	223	32·18	63·27	63·00	63·62	144·62

NOTE.—The depths of cuttings or embankments in the 2d column of the preceding Level-Book are found by calculation, or by carefully measuring them from the section by the vertical scale, but the latter method is not sufficiently correct. The computed half-widths in the 3d column are found by Problems I. and II. The corrected half-widths in columns 4th and 5th, by the five preceding Problems, according to the nature of the cuttings or embankments.

PROBLEM VI.

To find the quantity of land required for a projected railway.

CASE I.

In preparing the preliminary estimates for a projected railway, the quantity of land required for the purpose is usually found without paying any regard to the lateral inclination of the ground, by taking a considerable length of the section at once, especially if the surface thereof have a regular rise or fall, and by measuring the depth of the ends of such length with the vertical scale.

RULE.—Find the surface-widths, fences included, at each end of the given length, by Problems I. and II., add them together, multiply the sum by the length in chains, and divide the product by 1320 for the area in acres.

Ex. Let the length of the sectional surface be 18 chains, and the depths at the ends 22 and 38 feet, the bottom-width of the railway 33 feet, and the ratio of the slopes $1\frac{1}{2}$ to 1; required the area of the surface, the width of the side-fences being 9 feet each.

By Prob. II. $w + 2ar + 2f = 33 + (3 \times 22) + (2 \times 9) = 117$

$$w + 2br + 2f = 33 + (3 \times 38) + (2 \times 9) = 165$$

$$\begin{array}{r} 282 \\ 18 \\ \hline 1320 \begin{cases} 12 \\ 11 \end{cases} \left| \begin{array}{r} 507{\cdot}6 \\ \hline 42{\cdot}3 \end{array} \right. \\ \hline 3{\cdot}84545 \end{array} = \begin{array}{ccc} \text{A.} & \text{R.} & \text{P.} \\ 3 & 3 & 15 \end{array}$$

or by putting l = given length, and taking half the sum of the widths, there will result

$$\frac{(w + \overline{ra + b} + 2f)l}{660} = (33 + 1\tfrac{1}{2} \times \overline{22 + 38} + 2 \times 9)\ \frac{18}{660} = 3{\cdot}84545 \text{ acres.}$$

CASE II.

When the exact quantity of land for the railway is required.

RULE.—Take the whole widths at the end of every chain, from the 6th column of the Level-Book, for the several widths; add continually together the first and last widths, and twice the sum of all the intermediate widths, and divide the whole sum by 1320 for the area in acres.

Ex. Required the area corresponding to the several widths in the Level-Book at the end of Problem V.

$$\begin{array}{rl} 73{\cdot}55 & \\ 133{\cdot}03 & \\ 106{\cdot}82 & \\ 133{\cdot}93 & \\ 116{\cdot}84 & \\ 140{\cdot}29 & \\ \hline 704{\cdot}46 & \\ 2 & \\ \hline 1408{\cdot}92 & = \text{twice sum of intermediate widths} \\ 142{\cdot}30 & = \text{first width} \\ 144{\cdot}62 & = \text{last width} \\ \hline 1320 \begin{cases} 12 \\ 11 \end{cases} \left| \begin{array}{r} 169{\cdot}584 \\ \hline 14{\cdot}132 \end{array} \right. & \\ \hline 1{\cdot}28473 & = 1\text{A. } 1\text{R. } 5\tfrac{1}{2}\text{P.} = \text{area required.} \end{array}$$

NOTE.—It is very common in practice to find the areas of the quantities of land required from the several proprietors, by actual measurement from the two chain maps, made for the use of the contractors, after the several widths have been laid down thereon: copies being taken, at the same time, from the maps, on tracing-paper, showing the position and quantity of land required from each proprietor.

SECTION III.—CONTENTS OF CUTTINGS.

ON RAILWAY CUTTINGS IN GENERAL, AND TABLES FOR FINDING THEIR CONTENTS.

In preparing the preliminary estimates for a railway, the contents of the cuttings are usually found by tables for the purpose, the surface of the ground in the several cross-sections being assumed to be on a level with the centre of the line. But when power has been granted for constructing the line, the cross-sections are carefully taken at the end of every prominent variation of the surface of the ground, or, if consistent with accuracy, at the end of every one, two, or three chains in length; and the several cross-sections are then plotted on a large scale, which may be done by the methods given in the preceding Problems. Their areas may then be found by actual measurement; or they may be reduced, where the surface of the ground is uneven, to horizontal sections. The contents are found, either by taking the mean of every two succeeding sections, *which method is very erroneous where the areas of the sections differ greatly*, or by finding the contents from the tables, by using the mean depths of the several sections, *which method is correct; but the mean depth used in this method cannot be accurately found in many cases without considerable calculation.* Some use a mean of the mean depths as the bases for a mean area, *which method is also very inaccurate, especially where the areas of the extreme sections differ greatly.* The magnitude of the errors in both cases will be pointed out in the investigations at the end of these Problems.

On Earthwork Tables.

Tables for this purpose have been published by Sir John M'Neill, Mr Bidder, Mr Bashforth, Messrs Sibley and Rutherford, and others; all of which are well adapted for finding the contents of cuttings, assuming the surface of the ground to be laterally level with respect to the direction of the cutting. But none of these tables are accompanied with directions for finding the contents from sectional areas, *i.e.*, from the areas of working drawings, excepting Mr Bashforth's tables; *but his mathematical investigation of the rule for using them in finding the contents from working drawings, where the surface of the ground is laterally sloping or uneven, is founded on a false assumption, and therefore his results are erroneous, and especially so where the sectional areas differ considerably.* (See the investigations at the end of these Problems.)

THE GENERAL EARTHWORK TABLE.

(At the end of the book.)

This Table, with the help of the Auxiliary Earthwork Tables, Nos. 1 and 2, on the same sheet, possesses the advantage of being general for all varieties of slopes and bottom widths in common use, as well as for decimal parts of feet in the depths. It may also, with a very trifling preliminary calculation, be made to extend to every variety of bottom width and ratio of slope that can occur, if even the slopes of the two sides differ in the same cutting; and with the help of a table of square roots it will apply, with all attainable mathematical accuracy, to cuttings where the surface of the ground is uneven. The investigation of the method of forming the Tables and using them, will be given at the end of these Problems. The contents in the general Table, and those in Table No. 2, are calculated to the nearest unit for one chain in length, and checked by differences; the side-slopes being assumed to be extended till they intersect. The auxiliary Table No. 1 gives the depths of the intersections of the side slopes below the balance-line, and the corresponding number of cubic yards to be deducted from the contents for each chain in length.

PROBLEM I.

CASE I.

To find the contents of cuttings by the General Earthwork Table, and the Auxiliary Table No. 1, at the end of the book.

Let ABbdcC be a cutting, AB = *ab* = bottom width on the formation level, MM′ and *mm′* the perpendicular depths at the middle of the two ends of the cutting; AC, BD, *ac*, *bd*, the side-slopes, which, being prolonged two and two will meet at the points N and *n*; also MM′ and *mm′*, being prolonged, will meet at the same points. The distance M′N = *m′n* in feet and decimals is given in the Auxiliary Earthwork Table No. 1, for all bottom widths and ratios of slopes in common use, at which distance a line must be ruled on the section, parallel to the balance-line, or at the same distance + 2 feet from the line of the rails, in which latter case the balance-line need not be drawn. From the line thus ruled, the depths of the cutting must be measured to adapt them to the General Earthwork Table; or a mark might be made on the vertical scale with Indian ink (which is easily washed off) at the same distance, which mark might then be applied to the line of the

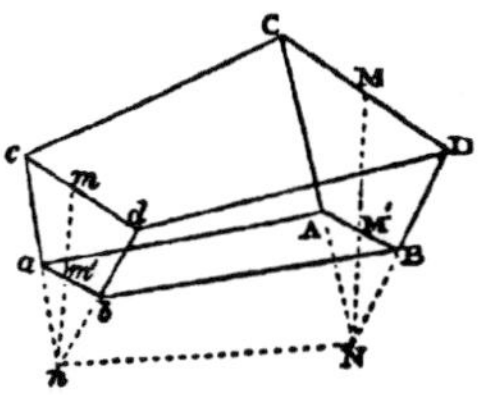

rails in measuring off the depths. For measuring the depths of embankments, the line must be ruled at the same distance—2 feet above the line of the rails. When the several quantities of a cutting or embankment have been taken from the Table, and their sum multiplied by the ratio of the slopes, the cubic yards to be deducted for each chain in length, for the particular bottom width and ratio of slopes, must be taken from Table No. 1 and multiplied by the whole length of the cutting, and the product, being subtracted from the result obtained from the General Table, will give the content of the cutting in cubic yards.

The method of using the Tables will best appear from the following examples.

Ex 1. Let the several depths of a railway cutting to the intersection of the slopes, at the end of every chain, be as in the annexed table, the bottom width 30 feet, and the ratio of the slopes $1\frac{1}{2}$ to 1; required the content of the cutting in cubic yards.

NOTE.—By the Table No. 1 the depth to be added to the depths of the cutting, for bottom width 30 ft. and ratio of slopes $1\frac{1}{2}$ to 1, is 10 ft., therefore the line from which the depths in the annexed table are measured is 10 + 2 = 12 ft. below the line of the rails. The corresponding number of cubic yards to be subtracted is carried to two places of decimals, or, if the nearest whole number had been taken, the quantity would have been in excess or defect by several cubic yards, when the cutting is of a considerable length.

Dist. in Chains.	Depths	Qts. per Table.
0	10	
1·00	29	1003
2·00	32	2276
3·00	33	2582
4·00	39	3175
5·00	35	3350
6·00	10	1365
For slope 1 to 1 . .		13751
" ½ to 1 . .		6875·5
" 1½ to 1 . .		20626·5
Subtract 366·67 × 6		= 2200
Content in cubic yds.		= 18426·5

Ex. 2. Let the several depths to the intersection of the slopes and their distances be as in the annexed table, the bottom width 30 feet, and the ratio of the slopes $1\frac{1}{2}$ to 1; required the content of the cutting.

NOTE.—When any of the distances is greater or less than 1 chain, the corresponding quantity must be multiplied by that distance; as in the cases of the distances between the depths 45 and 50, and between 30 and 10, the former distance being 3 chains and the latter 60 links.

Dist. in Chains.	Depths	Dist. greater than 1 Chain.	Qts. per Table.
0	10		
1·00	20		570
2·00	16		795
3·00	25		1044
4·00	32		1996
5·00	39		3091
6·00	45		4319
9·00	50	3 × 5520	16560
10·00	40		4971
11·00	30		3015
11·60	10	·60 × 1059	635·4
For slopes 1 to 1			36996·4
" ½ to 1			18498·2
" 1½ to 1			55494·6
366·67 × 11·60 . .			= 4253·4
Content in cubic yards . . .			51241·2

Ex. 3. The depth of a cutting to the intersection of the slopes and their distances in feet are as in the annexed table; the bottom width is 36 feet, and the ratio of the slopes 2 to 1; required the content of the cutting in cubic yards.

NOTE.—When the distances are given in feet, the quantities from the General Table must be multiplied by their respective distances; also the quantity from Table No. 1 must be multiplied by the whole distance, and the final result divided by 66, as in the annexed example. See Demonstration at the end of these Problems.

Dist. in Feet.	Depths	Qts. × by Length.	Products.
0	39		
100	61	6210 × 100	621000
188	50	7554 × 88	664752
178	37	4660 × 90	419400
For slopes 1 to 1			1705152
			2
,, 2 to 1			3410304
		396 × 278 =	110088
		66)	3300216
Content in cubic yards . .			50003

CASE II.

To find the contents of cuttings by the Tables, when the depths are given in feet and decimals of feet.

RULE.—Let any two succeeding depths be denoted by a and b, and let the decimal parts of the depths be respectively denoted by a' and b'; find the quantity corresponding to a and b from the General Table, as in the former case; then

$\frac{2a + b}{10}$, or its nearest whole number, and the decimal a' will show the number of cubic yards to be added in Table No. 2, and

$\frac{2b + a}{10}$, or its nearest whole number, and the decimal b' will show the cubic yards to be added in the same Table.

Ex. 1. Let the depths to the intersection of the slopes be 61·6 and 39·4 feet, their distance 1 chain, the bottom width 36 feet, and the ratio of slopes 2 to 1; required the content of the cutting in cubic yards.

$$\text{Put } 61 = a, \text{ its decimal } \cdot 6 = a'$$
$$39 = b, \text{ its decimal } \cdot 4 = b'.$$

Then the depths a and b, per General Table, give .	6210
$\frac{2a+b}{10}=16{\cdot}1$, or its nearest whole No. 16 and (a') ·6 per Table No. 2	78
$\frac{2b+a}{10}=13{\cdot}9$, or its nearest whole No. 14 and (b) ·4 per Table No. 2	46
For slopes 1 to 1 . . .	6334
	2
For slopes 2 to 1 . . .	12668
By Table No. 1, for bottom width 36 feet and ratio of slopes 2 to 1, No. of cubic yds. to be deducted .	396
Content in cubic yards	12272

Note 1. Care must be taken to use the decimal a' with $\frac{2a+}{10}$, and the decimal b' with $\frac{2b+a}{10}$, in finding the quantities in Table No. 2, as a mistake might easily be made in this matter, which would lead to an erroneous result.

2. If any two succeeding depths be nearly equal, and the sum of their decimal parts be together equal to 1 foot, or nearly so, by adding 1 foot to the lesser depth, and rejecting the decimal in the larger depth, and using the depths thus altered as whole nnmbers, a result sufficiently correct will be obtained; as in the following example.

Ex. 2. Let the depths of a cutting be 50·29 and 48·7 feet, their distance 1 chain, the bottom width 33 feet, and the ratio of the slopes $1\frac{1}{2}$ to 1; required the content.

By adding the decimal ·29, in the larger depth, to 48·7, the depths may be called 50 and 49, for which, by the General Table, the content is . . . . .

tent is	5990	for 1 to 1
	2995	for $\frac{1}{2}$ to 1
	8985	for $1\frac{1}{2}$ to 1
from which deduct, from Table No. 1	443·67	
	8541·33	cubic yards.

Note 1. The results of all the examples in the two preceding cases only differ by a very small fraction of a cubic yard from the true contents obtained by calculation from formula (1), page 456. The method of finding the contents to decimals or tenths of a foot in the depths has been particularly discussed, on account of its utility in finding the contents for actual contract-work from the working drawings, in which, as great accuracy is required, the contents should be found to two places of decimals, or to $\frac{1}{100}$ths of a foot in the depths, as in Case II. of the following Problem.

2. When one or both of the given depths exceed the limits of the table, find the content corresponding to half the two depths, and four times the result will be the content required.

PROBLEM II.

CASE I.

To find the content of a cutting between two cross-sections, the areas of which, in the intersection of the slopes, the length, the bottom width, and the ratio of the slopes, are given.

RULE.—Find the square roots of the given areas either by a table of square roots, or by actual extraction; with these roots, as depths, proceed to find the content from the General Table, as in Problem VII., from which deduct the quantity corresponding to the given bottom width and ratio of slopes from Table No. 1, and multiply the remainder by the length for the content.

NOTE. If the length be given in feet, multiply the content found by the above rule by the feet, and divide by 66 for the content.

Ex. 1. Let the areas of the two ends of a cutting be 5141 and 1444 square feet, the bottom width 30 feet, the length 1·60 chains and the ratio of the slopes $1\frac{1}{2}$ to 1; required the content of the cutting in cubic yards.

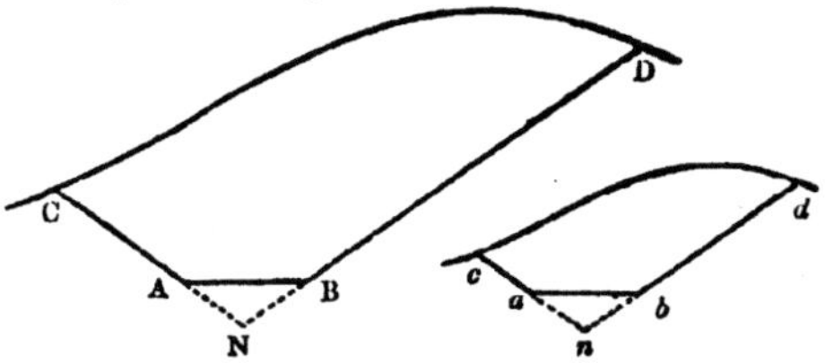

Let the annexed figures be the two cross-sections, the side slopes CA, DB; *ca*, *db*, being prolonged till they meet at N and *n*, the area CND = 5141, the area *cnd* = 1444; and since the bottom width AB = *ab*, and the ratio of the slopes are given, the solidity of the prism, the ends of which are ABN = *abn*, is given in Table No. 1, and is to be deducted from the content found by the General Table as in Problem VII.

The square roots of 5141 and 1444 are .	71·7 and 38
By General Table, for 71 and 38 . .	7483
By Table No. 2, for $\frac{(71 \times 2) + 38}{10} = 18$ and ·7 .	103
Content to intersection of slopes . .	7586
By Table No. 1, for bottom width 30, and slopes $1\frac{1}{2}$ to 1	366·67.
Content for 1 chain in length . .	7220·33
	1·60
Content for 1·60 chains in length . .	11552·528 cubic yards.

Case II.

When great accuracy is required, especially in the measurement of contract work, the second decimals, or $\frac{1}{100}$ths, of feet, must be considered in the calculation by taking for them $\frac{1}{10}$th of their corresponding quantities in Table No. 2.

Ex. Let the areas of the several cross-sections to the intersection of the slopes, and their distances, be as in the annexed table, the bottom width 36 feet, and the ratio of the slopes $1\frac{1}{2}$ to 1 ; required the content of the cutting in cubic yards.

Dist. in Chains.	Areas in Sq. ft.	$\sqrt{\text{Areas}}$	Product.	Contents.
0	2161	46·59		
1·00	3759	61·31	—	7147
2·00	5141	71·7	—	10832
4·20	4100	64·03	11271 × 2·20	24796
6·00	4221	64·97	10170 × 1·80	18306
8·00	3136	56·	8959 × 2·00	17918
9·00	2727	52·22	—	7160
				86159
From Table No. 1 for bottom width 36 ft. and slopes $1\frac{1}{2}$ to 1, 528 × 9·00				4752
Content in cubic yards . . .				81407

PROBLEM III.

To find the content of a railway cutting when the slopes of the two sides are different.

Case I.

When the depths are given.

Rule.—Find the content by the tables, as in the two preceding Problems, multiply the result by half the sum of the side slopes, and from the product deduct half the sum of the cubic yards, corresponding to the given bottom width and the ratios of the slopes of the two sides, multiplied by the whole length of the cutting, and the remainder is the content in cubic yards.

Ex. Let the depths and their distances be as in the annexed table, the bottom width 36 feet, and the ratios of the slopes of the two sides 1 to 1 and $1\frac{1}{2}$ to 1; required the content of the cutting.

Note.—Here the half sum of the first terms of the ratios of the slopes, viz. $\frac{1}{2}(1 + 1\frac{1}{2}) = 1\frac{1}{4}$ is the number by which the sum of the quantities, per tables, is to be multiplied, or $\frac{1}{4}$ added, as in the operation : and half the sum of the cubic yards, for the given bottom width and ratios of the two slopes, is multiplied by the whole length of the cutting, and deducted, agreeably to the rule.

Dist. in Chains.	Depths.	Qts. per Tables.
0	61	
1·00	52	7820
2·00	44·46	5473
For slopes 1 to 1		13293
,, $\frac{1}{4}$ to 1		3323·25
		16616·25
$\frac{1}{2}(528 + 792) \times 2$		1320
Content		15296·25

CASE II.

When the sectional areas are given.

RULE.—Find the content from the tables as in Problem VIII. and deduct for the quantity below the cutting, as in Case I. of this Problem.

NOTE.—This rule seems too obvious to require an example.

PROBLEM IV.

To apply the General Table and Table No. 2 to such bottom widths and ratios of slopes as are not found in Table No. 1.

Put w bottom width, and r to 1 the ratio of slopes.

Then $\frac{w}{2r}$ feet to be added to depth of cutting, or distance from bottom of cutting to the intersection of slopes.

And $\frac{11w^2}{18r}$ cubic yards to be deducted for each chain in length.

Ex. If the bottom width be 28 feet, and the ratio of slopes $1\frac{1}{4}$ to 1; then $w \div 2r$ $28 \div (2 \times 1\frac{1}{4})$ $11\frac{1}{5}$ 11·2 feet distance from bottom of cutting to intersection of slopes, and $11w^2 \div 18r$ $11 \times 28^2 \div 18 \times 1\frac{1}{4}$ 383·29 cubic yards to be deducted for each chain in length.

PROBLEM V.

To find the content of the cutting of a tunnel.

1. *When the length is given in yards, and the width and height in feet.*

RULE.—Multiply continually together the length, width, and mean height, and divide the product by 9.

2. *When the length is given in chains, and the width and height in feet.*

RULE.—Multiply the continued product of the length, width, and mean height by 22, and divide by 9.

Ex. The length of cutting of a tunnel is 1053 yards, its width 30 feet, and mean height 32 feet; required the content.

$$\frac{1053 \times 30 \times 32}{9} \quad 11232 \text{ cubic yards.}$$

The following example shows the magnitude of the errors of several methods, sometimes used, for calculating the contents of cuttings.

Ex. The areas of the two cross-sections of a cutting, to the intersection of the slopes, are 10324 and 400 square feet, their distance 4 chains, the bottom width 36 feet, and the ratio of the slopes 1 to 1; required the content of the cutting by the General Table, &c., and by the erroneous methods sometimes used.

Note.—The great difference of the sectional areas in this example shows very prominently the errors of several methods of finding the contents of cuttings: at the same time it is proper to remark that similar differences in the sectional areas very frequently occur in practice, and that the erroneous methods give the contents very near the truth *only* when these areas are nearly equal.

$\sqrt{10324}$	101·607	Content by General Tab., &c.	10394
$\sqrt{400}$	20		
Deduction from Tab. No. 1			792
Content for 1 chain in length			9602
			4
True content for 4 chains in length . . .			38408 cub. yds.

By Mr Bashforth's Method.

(1.) By Table No. 1 the depth below the formation-level to the meeting of the slopes is 18 ft.; hence the area of the triangle below it is $\frac{1}{2}$ (36 × 18) 324 square feet, which deducted from the given sectional areas, gives 10000 and 76 sq. ft. for the sectional areas, as used by Mr Bashforth.

Whence $\sqrt{10000}$	100	Content by Gen. Tab. &c.	8920·4
$\sqrt{76}$	8·718		
			4
Content for 4 chains in length			35681·6 c. y.

Error in defect, compared with the true content given above, *being above* $7\frac{1}{2}$ *per cent., by Mr B.'s method.* 2726·4

(2.) *By taking a mean of the areas for a mean section.*

$\frac{1}{2}$ (10000 + 76) × $\frac{22}{4}$ × 4 49260·4 cubic yards; *which exceeds the true content by* 10852·4 *cubic yards, being above* 22 *per cent. in excess.*

(3.) *By taking a mean depth the error in defect is just half of the preceding error, or nearly* 11 *per cent.*

OBSERVATIONS IMPORTANT TO THOSE CONCERNED IN THE CONSTRUCTION OF RAILWAYS, ON THE ERRONEOUS METHODS OF CALCULATING THE CONTENTS OF CUTTINGS, WHERE THE SURFACE OF THE GROUND IS UNEVEN; ALSO ON THE METHOD GIVEN IN THIS WORK.

The magnitude of the errors of the methods (1), (2), and (3), in the last example, for calculating the contents of cuttings, where the surface of the ground is uneven, from sectional areas, is strikingly apparent; and since many tables are not accompanied by any directions for their practical application in this particular case, one or other of the last two of these defective methods is still frequently used by engineers and contractors; thus causing continual disputes concerning the contents in consequence of the different parties using irreconcilable methods, some preferring one and some the other, as giving, in their respective judgments, the *true* content.

It is therefore the interest of those concerned in the construction of railways to adopt the method given in this work for finding the contents of cuttings, where the surface of the ground is laterally sloping, &c., as no other method combining all attainable mathematical accuracy has, to my knowledge, been yet published: those already published referring only to cuttings where the surface of the ground is either level, or the sections thereof assumed to be reduced to a level, which, in many cases, is a work of great labour to perform accurately. See the following investigations.

INVESTIGATION OF THE CONSTRUCTION AND USE OF THE GENERAL AND AUXILIARY TABLES, AND OF THE ERRORS OF OTHER METHODS.

Let ABDC, *abdc*, be vertical cross-sections of a railway cutting, the surface widths CD, *cd*, being horizontal, AC*ca*, BD*db*, the side-slopes, AB $= ab$ the bottom width. Prolong the planes of the

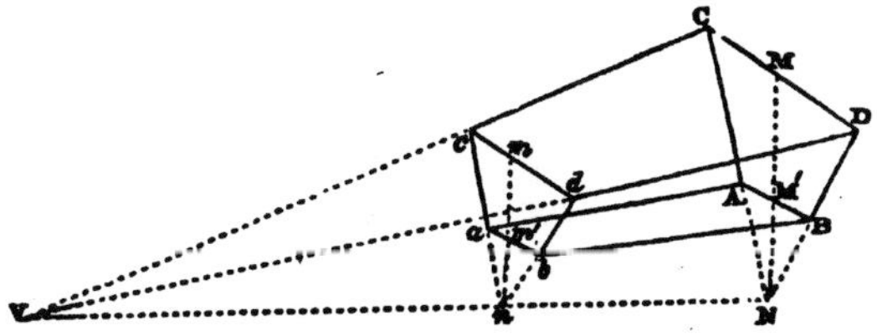

side slopes downwards and to the left till they meet in N*n*V, and also meet the prolongation of the plane CD*dc* in *c*V, *d*V. Because AB $= ab$, the cross-sections are ⊥ to VN. Let fall the ⊥ NM, *nm* on CD, *cd* respectively, bisecting them in M and *m*, and also bisecting AB, *ab* in M′ and *m*′. Put $MN = a$, $mn = b$, $Nn = Aa = Bb = l$, $AB = ab = w$, all in feet, and the ratio of the slopes, *i.e.*, CM : MN : : r : 1. Then

$\triangle$ CDN $= a^2r$, $\triangle\ cdn = b^2r$. By similar figures $a : b :: \text{NV} : \text{NV} - l$, whence $\text{NV} = \frac{al}{a-b}$,

$$\text{Pyramid VCND} = \tfrac{1}{3}\ \text{NV} \times a^2r.$$

$$\ldots\ldots\ Vcdn = \tfrac{1}{3}(\text{NV} - l)b^2r\text{; whence}$$

Frustum CDN$ncd = \frac{1}{3}\ \text{NV} \times a^2r - \frac{1}{3}\ (\text{NV} - l)\ b^2r = \frac{1}{3}r\ (\overline{a^2 - b^2}\text{NV} +\ b^2l) =$ (by substituting the value of NV)

$$\tfrac{1}{3}rl(a^2 + ab + b^2) \text{ cubic feet}$$

$$= \frac{rl}{81}(a^2 + ab + b^2) \text{ cubic yards} \quad . \quad . \quad . \quad . \quad (1)$$

In the General Table $r = 1$, and $l = 1$ chain $= 66$ feet; and $\therefore$ the solidity

$$S = \frac{22}{27}(a^2 + ab + b^2) \text{ cubic yards} \quad . \quad . \quad . \quad (2)$$

wherein a and b have all integral values from 0 to 72.

Auxiliary Table No. 1.

$$r : 1 :: \tfrac{1}{2}\ w(= \text{AM}') : \text{NM}' = \frac{w}{2r} \quad . \quad . \quad . \quad . \quad . \quad (3)$$

$$\text{whence } \triangle\ \text{ABN} = \triangle\ abn = \tfrac{1}{2}\ \text{AB} \times \text{NM}' = \frac{w^2}{4r} \quad . \quad . \quad . \quad . \quad . \quad (4)$$

$$\therefore \text{Prism ABN}nab = \frac{w^2l}{4r} \text{ sq. ft.} = \frac{w^2l}{108r} = \text{cubic yds.} \quad . \quad . \quad (5)$$

From (3) and (5) the depths to be added, and the cubic yards to be deducted in Table No. 1, are calculated, by taking $l = 66$ feet, and w and r all the values most commonly used in practice.

$$\therefore \frac{w^2l}{108r} = \frac{11w^2}{18r} = \text{cubic yards to be deducted, Table No. 1.} \quad . \quad (6)$$

$$\therefore \frac{11w^2\text{L}}{18r} = \text{cubic yards to be deducted for the length L.}$$

Let $\Sigma =$ sum of all the solidities s, s′, &c. (per General Table), the length of which is L, and $x = \frac{\text{H}w^2}{18r}$ from (6), then $\Sigma r =$ sum of solidities for slopes r to 1, from which deduct $x \times \text{L}$, and there results

$$\Sigma r - x\text{L} = \text{cubic yards in the whole cutting} \quad . \quad . \quad . \quad . \quad (7)$$

Whence the method of finding the contents of cutting in Problem VII.

Cor. 1. The content of a cutting having only two given depths is $(\text{S}r - x)\text{L}$ cubic yards. . . . . (8)

Cor. 2. If the ratio of the slopes of the two sides of a cutting be r to 1 and ρ to 1, and k, $\varkappa$ the corresponding cubic yards to be deducted. (Table No. 1.)

$$\tfrac{1}{2}\{\Sigma\,(r+\rho) - \mathrm{L}(k+\varkappa)\} \text{ cubic yards.}$$

Let in (2) a and b, hitherto supposed to be integral numbers, have the increments or decimals α and β, so that a and b become respectively $a+\alpha$ and $b+\beta$, which being substituted for a and b, neglecting the squares and product of α and β, as being comparatively small, there results :

$$\frac{22}{27}(a^2+ab+b^2)+\frac{22}{27}\,(2a+b)\,\alpha+\frac{22}{27}\,(2b+a)\,\beta \quad . \quad . \quad . \quad (9)$$

By subtracting (2) from (9), there results the sum of

$$\frac{22}{27}\,(2a+\ b)\alpha \text{ and } \frac{22}{27}\,(2b+a)\beta, \text{ or}$$

$$\frac{22}{27}\left(\frac{2a\ +\ b}{10}\right)10\alpha \text{ and } \frac{22}{27}\left(\frac{2b\ +\ a}{10}\right)10\beta \quad . \quad . \quad . \quad (10)$$

which are the increments of the formula (2) arising from the addition of the decimal parts α and β to the depths a and b, from which Table No. 2 has been calculated, the quantities $2a + b$, and $2b + a$ being divided by 10, to prevent a too great extension of the Table; and the corresponding factors α and β multiplied by the same number, that the product might retain their original value, the decimal points being still affixed to the values of α and β in the horizontal line at the top of the Table.

In consulting Table No. 2 the nearest whole numbers to $\dfrac{2a\ +\ b}{10}$ and $\dfrac{2b\ +\ a}{10}$ are taken as the small errors, thus resulting, will usually balance one another in a long cutting, and can never in any case amount to much.

Let the plane VCD revolve on the side VC, so that the lines CD, cd may incline from their hitherto assumed horizontal position, thus making the side slopes unequal; and let the area CDN A, and the area cdn B, the other symbols being the same as before; then

$$\sqrt{\mathrm{A}} : \sqrt{\mathrm{B}} :: \mathrm{VN} : \mathrm{VN} - l, \text{ whence VN } \quad \frac{l\sqrt{\mathrm{A}}}{\sqrt{\mathrm{A}} - \sqrt{\mathrm{B}}}.$$

Pyramid VCDN $= \frac{1}{3}$ VN $\times$ A, Pyramid Vcdn $= \frac{1}{3}\,(\mathrm{VN}-l) \times$ B, whence the frustum CDNncd = S = $\frac{1}{3}$ VN × A − $\frac{1}{3}$ (VN − l) × B = $\frac{1}{3}$ ($\overline{\text{A} - \text{B} \times \text{VN} + \text{B}l}$),

or $\text{S} = \frac{1}{3} l (\text{A} + \text{B} + \sqrt{\text{A} \times \text{B}})$, in cubic feet, and taking $l = 66$ feet

$= \frac{22}{27} (\text{A} + \text{B} + \sqrt{\text{A}} \times \sqrt{\text{B}})$ cubic yards . . . . . (11)

Therefore the Table must be consulted for depths $\sqrt{\text{A}}$ and $\sqrt{\text{B}}$, the ratio of the slopes being included; for if a and b be the mean depths for the areas A and B respectively, the slopes being r to 1, then $a^2 r = \text{A}$, $b^2 r = \text{B}$ and $\frac{22}{27} (\text{A} + \text{B} + \sqrt{\text{A}} \times \sqrt{\text{B}}) = \frac{22r}{27} (a^2 + ab + b^2)$.

Therefore in reality the Table is consulted for the depths $a \sqrt{r} = \sqrt{\text{A}}$, and $b \sqrt{r} = \sqrt{\text{B}}$.

If Σ' = sum of all the solidities S, S′, &c., for depths $\sqrt{\text{A}}$ and $\sqrt{\text{B}}$, &c., and length L; then by subtracting xL, as in (7), there results $\Sigma' - x$L cubic yards in the whole cutting . . . . (12)

Whence the Rule in Problem VIII.

Cor. 1. The content of a cutting, having only two given sectional areas, is (S − x) L cubic yards . . . . (13)

Cor. 2. Formulæ (11), (12), and (13) will evidently hold, if, instead of the straight lines CD, cd, being the surface edges of the cross-sections CDN, cdn, of the cutting, these lines be the chords of similar curves forming the surface edges of the cross-sections, and the similarly situated points in the two curves be joined by right lines (which prolonged will all meet in the vertex V), thus forming the surface of the cutting, a necessary condition in taking cross-sections; or the small inequalities of the earth's surface between the cross-sections must be balanced, as nearly as can be judged by the eye, so that the surface may fulfil this condition, in order that all attainable mathematical accuracy may be arrived at in finding the contents.

THE ERRORS OF OTHER METHODS OF FINDING THE CONTENTS OF RAILWAY CUTTINGS.

Mr Bashforth, in his investigations for finding the contents of railway cuttings, where the surface of the ground is level, or assumed to be so, includes the prism ABNnba, and afterwards deducts it, which is mathematically accurate. But in finding the contents from sectional areas, where the surface of the ground is laterally sloping, or uneven, he altogether leaves out the above-named prism; for he says (Art. 15, p. 11 of his work), "The only way to proceed in such a case is—find the areas of the cross-sections in square feet, take out of Barlow's Tables the square root of each, and treat these square roots precisely as if they had been measured heights, excepting that there will be nothing to be deducted for the prism, and no multiplication for the slopes, these having been already accounted for in finding the areas."

Let A' and B' be the sectional areas CDBA, *cdba* respectively, as used by Mr B., and $l = Nn =$ length of the cutting (fig. page 452), then according to his method, the content of the cutting is

$$\frac{l}{81}\left(A' + \sqrt{A' \times B'} + B'\right) \quad . \quad . \quad . \quad . \quad . \quad . \quad (14)$$

which is erroneous in every case, except where the areas represented by A' and B' are similar and equal. For when the equal areas ABN, *abn* are omitted, the areas represented by A' and B' are dissimilar, and $\therefore$ the proportion $\sqrt{A'} : \sqrt{B'} :: NV : nV$, on which Mr B.'s rule is founded, does not hold, it being true only in respect to the areas CDN, *cdn*, because the solids VCDN, V*cdn* are similar. Moreever, if the plane VCD revolve on the side CD so as to bring the lines *cd*, *ab*, to coincidence, or almost to coincidence, in the points *b* and *d*, the dissimilarity of the planes CDBA, *cdba* will become more strikingly evident, whereas with the addition of the Δs ABN, *abn*, they still remain similar.

Let $\Delta ABN = abn = \alpha$, then $A' + \alpha$ and $B' + \alpha$ are the sectional areas corresponding to A and B in formula (11) of my investigation, whence by that formula the true content, including the prism ABN*nba*, is

$$\frac{l}{81}\left(A' + B' + 2\alpha + \sqrt{\overline{A' + \alpha} \times \overline{B' + \alpha}}\right) \text{ cubic yards.}$$

But the solidity of the prism is $\frac{al}{27} = \frac{3al}{81}$ cubic yards.

Whence the content of the prismoid or cutting is

$$\frac{l}{81}\left(A' + B' + 2\alpha + \sqrt{\overline{A' + \alpha} \times \overline{B' + \alpha}} - 3\alpha\right), \text{ or}$$

$$\frac{l}{81}\left(A' + B' + \sqrt{\overline{A' + \alpha} \times \overline{B' + \alpha}} - \alpha\right),$$

from which subtract the content according to Mr B. (14), and there results.

$$\frac{l}{81}\sqrt{\overline{A' + \alpha} \times \overline{B' + \alpha}} - \sqrt{A' \times B'} - \alpha) = \text{error in defect.}$$

Let $A' = \frac{1}{4}a$, and $B' = 36a$, which is a case that may occur in practice, then

$$\frac{l}{81}(\sqrt{A' + a \times B' + a} - \sqrt{A' \times B'} - a) = \frac{al}{81}(\tfrac{1}{2}\sqrt{185} - 3 - 1)$$

$= \frac{al}{81} \times 2{\cdot}8007 =$ error in defect by Mr B.'s method.

and $\frac{l}{81}(A' + \sqrt{A' \times B'} + B') = \frac{al}{81}(\frac{1}{4} + 3 + 36) = \frac{al}{81} \times 39\frac{1}{4} =$ whole content by Mr B.'s method.

$\therefore$ whole content : error in defect : : $\frac{al}{81} \times 39\frac{1}{4} : \frac{al}{81} \times 2{\cdot}8007$

$39\frac{1}{4}$: 2·8007

100 : $7\frac{1}{9}$, or $7\frac{1}{9}$ *per cent.*

the error in defect . . . . . (15)

See Ex. at the end of Problem X., where the error of Mr Bashforth's method is shown by giving actual areas.

NOTE.—The errors of this method are not so prominent where the sectional areas approach near to equality, as in the case of the Burnley cross-sections, in the Ex. Art. 20, page 13, of Mr B's work: his method, however, is erroneous to a greater or lesser extent, in every case, except where the sectional areas are equal and similar; for they cannot be similar without being equal, while the bottom width remains the same.

Mr Bashforth says, in defence of his method of finding the contents of cuttings from sectional areas, in the "Mechanics' Magazine" for Sept. 11, 1847, p. 249, "In the case of contract estimates, numerous cross-sections ought to be taken; and it matters little whether we start from the intersection of the slopes or the formation-level." *The above-noticed error* (15) *shows that it "matters" so much as* $7\frac{1}{9}$ *per cent. in defect.* Besides, there is no need of taking "numerous cross-sections," as Mr B. recommends, especially where the surface of the ground is laterally sloping like a geometrical plane, or curved like a conical surface, one or other of which cases very frequently occurs, so that cross-sections, taken at a considerable distance from one another to the intersection of the slopes, may be considered as similar, or so very nearly so as not to induce any important mathematical error, which conditions may be easily determined by the eye. Moreover, the expense and trouble of taking "numerous cross-sections," plotting them, and finding their areas, are very considerable; and, therefore, ought to be avoided, together with all erroneous methods, such as Mr B.'s, of finding the contents of cuttings, as his "numerous cross-sections" only tend to diminish the errors, without wholly getting rid of them.

The error of the method of finding contents by mean areas.

Let A and B be the areas of two cross-sections of a cutting to the intersection of the slopes, and l its length: then the mean area is $\frac{1}{2}$ (A + B), and the content in cubic yards is $\frac{1}{27} l \times \frac{1}{2} (A + B)$ $\frac{1}{81} l \times \frac{3}{2} (A + B)$; from which subtract the true content, equation (11), p. 455, and there results

$\frac{1}{81} l \{\frac{3}{2} (A + B) - (A + B + \sqrt{AB})\} = \frac{1}{162} l (\sqrt{A} - \sqrt{B})$ = *error in excess.*

Which error is very great when the areas A and B differ considerably. See (2), p. 451.

The error of the method of finding the content by mean depth.

Let a and b be the depths of two cross-sections to the intersection of the slopes, the ratio of which is r to 1, and l = length of the cutting; then $\frac{1}{2} (a + b)$ = mean depth, $\frac{1}{4} r (a + b)^2$ = mean area, and the content in cubic yards $= \frac{1}{27} l \times \frac{1}{4} r (a + b)^2 = \frac{1}{81} rl \times \frac{3}{4} (a + b)^2$;

which subtract from equation (1), p. 411, and there results

$\frac{1}{81} rl \{a^2 + ab + b^2 - \frac{3}{4} (a + b)^2\} = \frac{1}{324} rl (a - b)^2$ = error in defect:

Which error is very considerable, when the depths a and b differ greatly.

NOTE.—The errors of these methods, as here shown, are the same as if the areas and depths had only extended to the formation-level, since a common quantity, *i.e.* the prism below the formation-level, is here included, and afterwards excluded, by taking the differences.

SECTION IV.

TUNNELLING.

(1.) Previous to setting out the earthwork of a tunnel, the levelling operation must be repeated with great care, and should also be checked by the method given in Art. (22), Section I., especially if the tunnel pass under a very high summit: for, if the section be incorrect, the gradient or gradients, on which the tunnel is formed, will not meet at the points shown thereon, and thus embarrass the mining operation.

(2.) If the tunnel is formed on one gradient, as BD, Plate XIII.

the gradient must incline to one of the extremities of the tunnel, as at D, in order to discharge the water generated therein. Strong poles or masts must be firmly fixed on the surface, in the intended direction of the tunnel, of which one must be on the summit of the hill; at which place a temporary observatory is frequently erected, especially if the summit be a very high one, and the tunnel a very long one. Shafts must be sunk at the distance of four or five chains from one another, in the direction of the poles (and observatory, if there be one), in order to ventilate the tunnel, as well as to check the accuracy of the work as it proceeds. If the tunnel be a long one, it would be preferable, if convenient, to form it on two gradients, inclining to its opposite extremities to liberate the water, and thus to aid the mining operation, which is commonly commenced at both ends of the tunnel at the same time.

(3.) When it is necessary to have a curve in the direction of a part or of the whole of the tunnel, that direction must be carefully laid down on the surface, by the methods given in Section III., making allowance for acclivities and declivities, poles or masts being fixed therein, as pointed out in Art. (2), that the shafts may be sunk so as to meet the mining operations of the tunnel, as well as to check their accuracy in point of direction, and this will be the more especially necessary in the curved part of the tunnel.

(4.) The mining operation of the tunnel should commence when the depth of the cuttings at each end is about 60 feet. The width and depth of the excavation of a tunnel, on the narrow gauge, should be about 30 feet each, and must be dug 5 or 6 feet below the intended line of the rails, to give space for the inverted arch and the ballasting, excepting where the excavation is made through rock sufficiently hard to form the side-walls of the tunnel, in which case 22 or 24 feet in width, and about 26 feet in height, will be sufficient, the excavation in this case being terminated below by the balance-line, or formation-level. The depth and width of the excavation for a tunnel on the broad gauge must, in both cases, be proportionately larger.

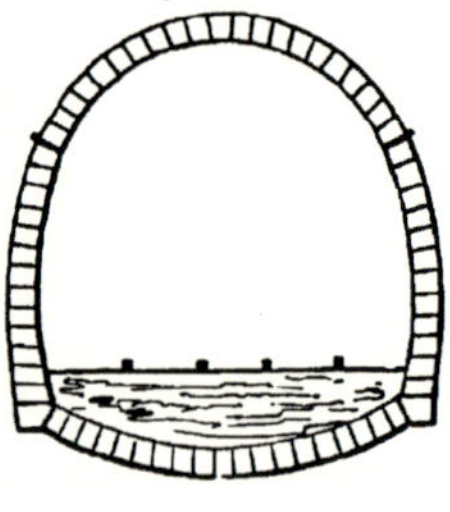

The annexed figure is a cross-section of the masonry of a tunnel, which, of course, is such as is required where the tunnel is made through loose earth, only the arch above being required when made through hard rock.

NOTE.—The remarks at Articles (25) and (27), Section II., refer only to the projection of tunnels on parliamentary maps.

A TABLE OF THE DIMENSIONS OF SEVERAL EXISTING TUNNELS.

Names of Railways and of Tunnels.		Length in Yards.	Width in Feet.	Height in Feet.	Shafts.
London and Birmingham	Primrose Hill	1120	22	22	5
	Weedon . . .	418			
	Kilsby . . .	2398			10
Cromford and High Peak—Buxton		580	21	16	
Great Western—The Box Tunnel		3227	35	29	13
Manchester and Leeds	Littleborough .	2869	24	21½	
North Midland	Claycross . .	1806	22	21	
	Melford . . .	836			
London and Brighton	Merstham	1780			
	Balcombe	1192			
Sheffield and Manchester	Five tunnels Total length	6245			
Chester and Holyhead	Three tunnels Total length	2130			
South-Western	Seven tunnels Total length	1998			

SECTION V.

VIADUCTS, AQUEDUCTS, SKEW ARCHES, ETC.

What are here given on these subjects are necessarily only outlines, referring to some few existing structures, without entering into the details of specifications, working drawings, &c., which are foreign to the nature of this work, and properly belong to the department of the architect.

Railway Viaducts.—It would be impossible to enumerate the whole of the viaducts constructed since the introduction of railways. Structures in timber, brick, stone, and iron, of various designs, have been erected; and in some cases there is a novelty of principle accompanied with great boldness of execution.

The Brick Viaduct at Maidenhead, constructed by Mr Brunel for the Great Western Railway, is one of the best examples in that material: it is composed of two elliptical arches spanning the Thames, each 128 feet, with a versed sine of 24¼ feet; the pier between the two arches is 30 feet in width. The arch in the middle is 5¼ feet in thickness, which gradually increases towards the abutments. Besides these two grand brick arches, on each bank of the river are four others; those on the abutments span 21 feet each, and the six others 28 feet each.

The Viaduct over the Ouse, near York, consists of three arches, each 66 feet span; the piers are 10 feet in thickness, and the width of

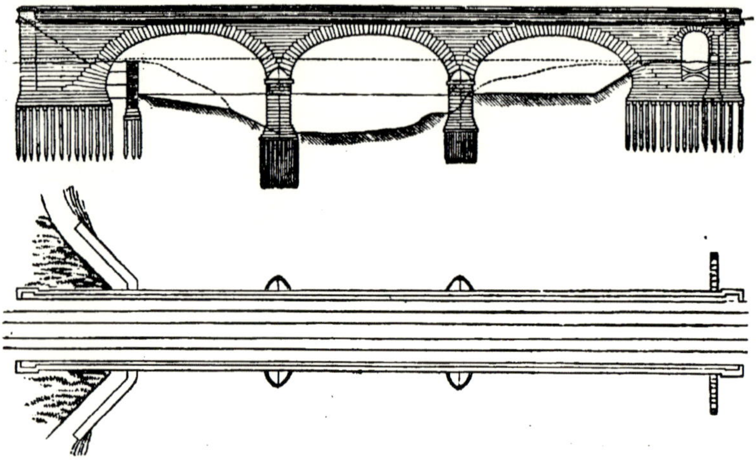

the arches 28½ feet on the soffite, and their thickness at the keystone 3½ feet, the voussoirs gradually increasing towards the springing.

This is a specimen of a stone viaduct combining great strength and elegance in its construction.

Aqueduct of Pont-y Cysyllte is 4 miles from Chirk, over the river Dee; its length is 1007 feet, and the river is 127 feet below the water-level of the canal carried over it.

To construct an aqueduct upon the usual principles, with piers and arches above 100 feet in height, and of a sufficient breadth and strength to afford room for a puddled water-way, would have been not only expensive but extremely hazardous. Telford, who had already carried the Shrewsbury canal by a cast-iron trough 16 feet above the level of the ground, formed the idea of doing the same in the present instance, which was approved and finally adopted. The foundation on which the piers are erected is a hard sandstone; their height above low water in the river is 121 feet: at the botton they are 20 feet by 12, at the top 13 feet by 7½. For a height of 70 feet above the foundations, they are built solid, and the remaining 50 feet hollow, the walls being only 2 feet in thickness, with one cross inner wall; by this means the centre of gravity of the piers is thrown lower, and the masonry economised. The width of the water-way is 12 feet; of which the towing-path covers $4\frac{1}{2}$ feet, leaving $7\frac{1}{2}$ feet for the boat: as the towing-path is supported by iron pillars, the water fluctuates, and recedes freely as the boat passes.

There are 18 of these stone piers, besides the abutments; the span of the arches is 45 feet, and their rise above the springing $7\frac{1}{2}$ feet, and the total expense of its construction was 47,018*l.*

This aqueduct almost rivals the works of a similar kind left us by the ancient Romans: the introduction of iron, however, for a water-course, is a novelty with which they were unacquainted: in this instance it has proved admirably well fitted for the purpose to which it is applied: had the channel been constructed with stone or brick at this great elevation, it would have been less secure, as there would have been a constant danger of leakage.

SKEW OR OBLIQUE BRIDGES.

One of the skew bridges on the *London and Birmingham line*, a fine erection of this kind, is $23\frac{1}{2}$ feet in height from the surface of the road below to that of the rails, making an angle of 32° with the road: the direct span of the arch is 21 feet, and oblique span 40. The arch, which is $2\frac{1}{2}$ feet thick, is the segment of a cylinder, the internal radius of which is $12\frac{1}{2}$ feet, and the versed sine $5\frac{2}{3}$ feet. The angle of which the coursing joints of the soffite cross the axis of the cylinder is 53° 25′, and the joints of the face of the arch all converge to a point $32\frac{1}{2}$ feet below the axis of the cylinder, and 45 feet below the crown of the arch.

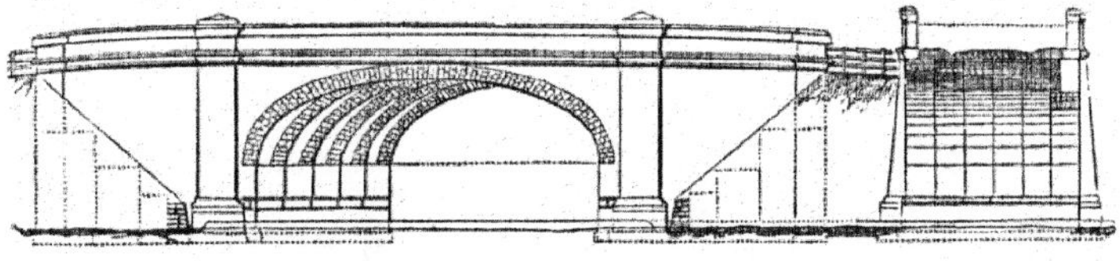

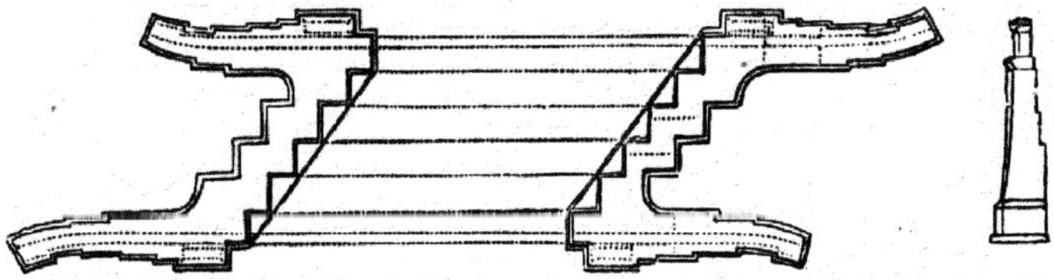

The Midland Counties Railway possesses another variety of skewed brick bridge, the span of which is $42\frac{1}{2}$ feet, and versed sine 11 feet;

the arch consists of six ribs, 2 bricks in depth, and 4 feet in thickness, the total breadth of the bridge, measured at right angles to the face, being 24 feet.

Manchester and Birmingham Railway.—In Fairfield Street is a skew bridge, the oblique span of which is 128¾ feet, with a versed sine of 12 feet; the width from face to face is 31 feet.

Six ribs of iron abut on as many independent walls, and project before each other 13 feet; an ornamental stone parapet and cornice crown this viaduct, presenting a novel and agreeable appearance.

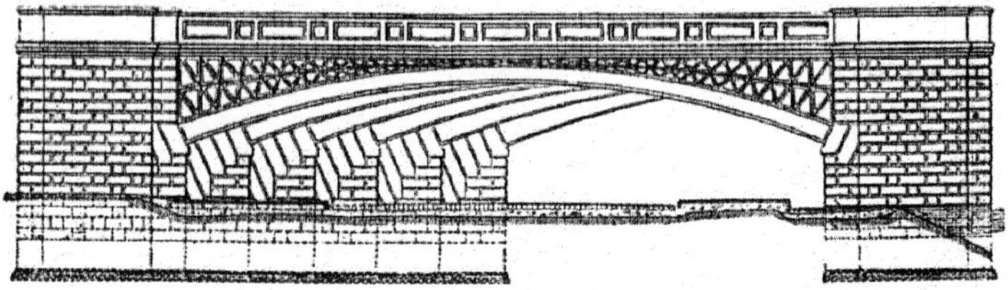

This bridge, executed after the designs of Mr. G. W. Buck, is remarkable for its acute angle, which is 24½°. The weight of the iron-work employed on the six ribs was 540 tons, and the whole was admirably secured together; the remainder of the viaduct is formed with brick arches of 45 feet span.

DIMENSIONS OF VIADUCTS IN VARIOUS RAILWAYS.

In the Grand Junction Railway, at Vale Royal, is a viaduct of stone over the river Weaver, in which are five arches, each 63 feet in span and 60 in height; the length of the viaduct is 456 feet. Where this railway crosses the Mersey and Irwell Canal there is a viaduct of stone having 12 arches; the two in the centre are 75 feet in span, and the remainder from 40 feet to 12½.

In the Newcastle and Carlisle Railway, near Brampton, is a viaduct, which crosses the public road and the river Gelt, at the height of 80 feet above the bed of the river, over which it is carried in an oblique direction. The arches, which are three in number, are 33 feet in direct span, and are built at an angle of 45°.

In the Birmingham and Derby Junction Line, between Kingsbury and Tamworth, over the Anker, is a viaduct of 18 arches, each of 30 feet span, and one oblique arch of 60 feet span; its height above the river is 23 feet, and the cost was 18,000*l.* *In the same line,* between Tamworth and Burton-on-Trent, is a viaduct ¼ mile in length, built upon above 1000 piles, driven 15 feet below the bed of the river.

In the Newcastle and Shields Railway there is a viaduct over the Ouseburn of 9 arches, two of which at the ends are of stone, and the others are of timber, resting on stone piers; the three central arches have each a span of 116 feet, and two others 110; the total length of the viaduct is 750 feet, and its height above the water 180 feet. *In the same line,* the viaduct at Willingdon Dean has 7 timber arches, five of which span 120 feet each, and the two exterior each 115 feet; the whole length is 1050 feet, and the height 82 feet.

In the Taff Vale Railway, near Quaker's Yard, is a viaduct crossing the Taff, the length of which is 600 feet, and the height above the river 100 feet, having 6 arches. *In the same line*, at the conflux of the Rhondda and Taff, is a viaduct having an arch of 100 feet span, and 60 feet in height.

In the London and Brighton Railway, across the valley of the Ouse, is a viaduct 1437 feet in length, and the height varies from 40 to 96 feet; it is formed of 37 brick arches of 30 feet span.

The London and Greenwich Railway is a continuous viaduct of more than 1000 brick arches, each 18 feet span, 22 feet in height, and 25 feet in width. It is 3¾ miles in length, and cost 266,322*l.* per mile.

The extension of the South-Western Railway through the metropolis, from the Nine Elms to Waterloo Bridge, is a continuous viaduct, like the last-mentioned one, in which are several strong and elegant oblique iron arches for the purpose of crossing some of the principal streets.

The student's attention should also be directed to the many magnificent railway bridges and viaducts of *iron* which have been erected of late years; such as the *Britannia and Conway Bridges* in North Wales; the *High Level Bridge* at Newcastle-on-Tyne; and the great *Victoria Bridge* in Canada by Mr Robert Stephenson; the *Windsor*, *Chepstow*, and *Saltash* bridges by Mr Brunel; the new *Charing Cross Bridge* by Mr Hawkshaw; and the *Viaduct at Crumlin* in Monmouthshire, by Messrs Kennard. Descriptions of these will be found in engineering works.

I shall conclude this section by recommending to those who wish for scientific and practical information of the first order on this subject, "The Theory, Practice, and Architecture of Bridges," by J. Hann and others; "Practical and Theoretical Essay on Oblique Bridges," by G. W. Buck, M. Inst. C.E.; "Treatise on the Equilibrium of Arches," by Joseph Gwilt, architect, F.S.A.; Cresy's "Encyclopædia;" and a "Practical Treatise on the Construction of Oblique Arches," by J. Hart.

SECTION VI.

SUPERELEVATION OF EXTERIOR RAIL IN CURVES.

The superelevation of the exterior rail in curves, the radii of which are within certain limits, is absolutely necessary to counteract the centrifugal force caused by the velocity of the train, since all moving bodies have a tendency to continue their motion in a direct line. From this cause the carriages of a railway train are driven towards the exterior rail, and would finally be thrown off the rails, were it not for the conical inclination of the tire and the flanges of the wheels.

Let W = weight of the moving body or train, V = its velocity per second, R = radius of the curve, and g = force of gravity at the earth's surface ; then, by Dynamics, the centrifugal force

$$f = \frac{WV^2}{gR} \quad . \quad . \quad . \quad . \quad . \quad (1)$$

When R = 1 mile = 5280 feet, V = velocity = 60 miles per hour = 88 feet per second, and $g = 32\frac{1}{6}$ feet, then .

$$f = \frac{W \times 88^2}{32\frac{1}{6} \times 5280} = \tfrac{1}{22}W;$$

that is, the centrifugal force that urges the moving body to leave the curve, in this case, is $\frac{1}{22}$ of its weight.

This force is slightly counteracted by the conical inclination of the tire of the wheels, each pair of which are firmly fixed on the axle that turns with them. This inclination of the tire, together with the lateral play of the flanges of $\frac{1}{2}$ an inch on each side, and the centrifugal force impelling the carriages of the train, when moving in a curve, towards the exterior rail, enlarge the diameter of the exterior wheel, and diminish that of the interior, thus causing the train to roll on conical surfaces, which necessarily produces a centripetal force, the centre of which force is the vertex of the cone, of which the increased and diminished diameters of each pair of wheels are sections.

Let d be the outer diameter of the wheels, δ the increment and consequently the decrement that the diameters of the exterior and interior wheels respectively receive, through the joint action of the centrifugal force and the inclination of the tire: then under these circumstances the respective diameters of the exterior and interior wheels will be

$$d \times \delta \text{ and } d - \delta$$

also if R′ be the radius of a circle which the centre of the carriage would describe in consequence of the inclination of the tire of the wheels, and b the breadth of the road or gauge : then $R' + \frac{1}{2}b$ and $R' - \frac{1}{2}b$ are the radii which would be respectively described by the exterior and interior wheels ; and by similar triangles

$$d + \delta : d - \delta :: R' + \tfrac{1}{2}b : R' - \tfrac{1}{2}b,$$

whence $\quad d : \delta :: 2R' : b$, and

$$R = \frac{bd}{2}$$

Or, if $\frac{1}{n}$ = inclination of the tire, and Δ the deviation of the wheels, the increment or decrement $\delta = \frac{2\Delta}{n}$; and, by substitution,

$$R = \frac{bdn}{4\Delta} \quad . \quad . \quad . \quad . \quad (2)$$

Also W and V representing the weight and velocity of the train, as in (1), and g the force of gravity, the centripetal force corresponding to the radius R′ will be

$$f' = \frac{WV^2}{gR'} \quad . \quad . \quad . \quad . \quad (3)$$

Or, by substituting the value of R′ from (2)

$$f' = \frac{WV^2}{g} \times \frac{4\Delta}{bdn} \quad . \quad . \quad . \quad . \quad (4)$$

Since the forces f and f' (1) and (3), act in contrary directions, they will hold each other in equilibrium when they become equal, and the train will cease to have a tendency to leave the curve; this takes place when

$$\frac{WV^2}{gR} = \frac{WV^2}{gR'}$$

$$\text{or } R = R'$$

Also from (1) and (4)

$$\frac{WV^2}{gR} = \frac{WV}{g} \times \frac{4\Delta}{bdn}$$

whence

$$\Delta = \frac{bdn}{4R} \quad . \quad . \quad . \quad . \quad . \quad (5)$$

which is the deviation required to produce the equilibrium between the centripetal and centrifugal forces of the train. Therefore since R = R′, *i.e.*, the vertex of the imaginary cone of which each pair of wheels are sections will coincide with the centre of the curve, there will in consequence be no dragging of either of the wheels on the rail.

In practice, however, it is safer to neglect the effect of the coning of the tires, and to calculate the superelevation of the rail as if the tires were cylindrical.

The rule for this is as follows:—

Let R = radius of curve in chains
b = breadth of gauge of the line in feet; or more properly, the distance from centre to centre of rails.

V = velocity of the train in miles per hour.

x = superelevation to be given to the outside rail in inches.

Then

$$x = \frac{b\text{V}^2}{82{\cdot}5\ \text{R}}$$

For the narrow gauge, where $b = 4$ ft. 11 in. $= 4{\cdot}91$,

$$x = \frac{\text{V}^2}{17\text{R}} \text{ nearly.}$$

Example.—On a curve ½ mile (= 40 chains) radius it is expected that trains will travel at 40 miles an hour. What ought to be the superelevation of the outside rail?

$$x = \frac{40^2}{17 \times 40} = \frac{1600}{680} = 2{\cdot}35 \text{ inches.}$$ *Answer.*

TABLES FOR REDUCTION OF DISTANCES AND CORRECTION OF LEVELS FOR CURVATURE, ETC.

No. 1. Reduction for each Chain for the following Angles of Elevation.

Angles.		Reduction in Links.
3°	0′	0·137
3	30	0·187
4	0	0·244
4	30	0·308
5	0	0·381
5	30	0·460
6	0	0·548
6	30	0·643
7	0	0·745
7	30	0·856
8	0	0·973
8	30	1·098
9	0	1·231
9	30	1·371
10	0	1·519
10	30	1·675
11	0	1·837
11	30	2·008
12	0	2·185
12	30	2·370
13	0	2·553
13	30	2·763
14	0	2·970
14	30	3·185
15	0	3·407
15	30	3·637
16	0	3·874
16	30	4·118
17	0	4·370
17	30	4·628
18	0	4·894
18	30	5·168
19	0	5·448
19	30	5·736
20	0	6·031
20	30	6·333
21	0	6·642
21	30	6·958
22	0	7·282
22	30	7·612
23	0	7·950
23	30	8·294
24	0	8·645
24	30	9·004
25	0	9·369
25	30	9·741
26	0	10·121
26	30	10·507
27	0	10·899
27	30	11·299
28	0	11·705
28	30	12·118
29	0	12·538
29	30	12·964
30	0	13·397

No. 2. Difference between Apparent and True Level for distances in Chains. Correction in Decimals of Feet.

Distances in Chains.	For Curvature.	For Refraction.	For Curvature and Refraction.
3	·0009	·0001	·0008
3½	·0013	·0002	·0011
4	·0017	·0002	·0015
4½	·0021	·0003	·0018
5	·0026	·0004	·0022
5½	·0031	·0004	·0027
6	·0037	·0005	·0032
6½	·0044	·0006	·0038
7	·0051	·0007	·0044
7½	·0058	·0008	·0050
8	·0067	·0010	·0057
8½	·0075	·0011	·0064
9	·0084	·0012	·0072
9½	·0094	·0013	·0081
10	·0104	·0015	·0089
10½	·0115	·0016	·0099
11	·0126	·0018	·0108
11½	·0138	·0020	·0118
12	·0150	·0021	·0129
12½	·0163	·0023	·0140
13	·0176	·0025	·0151
13½	·0190	·0027	·0163
14	·0204	·0029	·0175
14½	·0219	·0031	·0188
15	·0234	·0033	·0201
15½	·0250	·0036	·0214
16	·0267	·0038	·0229
16½	·0284	·0041	·0243
17	·0301	·0043	·0258
17½	·0319	·0046	·0273
18	·0338	·0048	·0290
18½	·0357	·0051	·0306
19	·0376	·0054	·0322
19½	·0396	·0056	·0340
20	·0417	·0060	·0357
20½	·0438	·0063	·0375
21	·0459	·0066	·0393
21½	·0481	·0069	·0412
22	·0504	·0072	·0432
22½	·0527	·0075	·0452
23	·0551	·0079	·0472
23½	·0575	·0082	·0493
24	·0600	·0086	·0514
24½	·0625	·0089	·0536
25	·0651	·0093	·0558
25½	·0677	·0097	·0580
26	·0704	·0100	·0604
26½	·0731	·0104	·0627
27	·0759	·0108	·0651
27½	·0788	·0113	·0675
28	·0817	·0117	·0700
28½	·0846	·0121	·0725
29	·0876	·0125	·0751
29½	·0906	·0129	·0777
30	·0937	·0134	·0803

No. 3. Difference between Apparent and True Level for distances in Miles. Correction in Feet and Decimals.

Distances in Miles.	For Curvature.	For Refraction.	For Curvature and Refraction.
¼	·042	·006	·036
½	·167	·024	·143
¾	·375	·054	·321
1	·667	·095	·572
1½	1·501	·215	1·286
2	2·668	·381	2·287
2½	4·169	·596	3·573
3	6·003	·856	5·147
3½	8·171	1·167	7·004
4	10·672	1·525	9·147
4½	13·547	1·930	11·517
5	16·675	2·382	14·293
5½	20·177	2·882	17·295
6	24·012	3·430	20·582
6½	28·181	4·026	24·155
7	32·683	4·669	28·014
7½	37·519	5·360	32·159
8	42·688	6·098	36·590
8½	48·191	6·884	41·307
9	54·027	7·718	46·309
9½	60·197	8·600	51·597
10	66·700	9·529	57·171
10½	71·520	10·217	61·303
11	80·687	11·527	69·160
11½	88·167	12·594	75·573
12	96·000	13·714	82·286
12½	104·167	14·881	89·286
13	112·667	16·095	96·572
13½	121·500	17·357	104·143
14	130·667	18·666	112·001
14½	140·167	20·024	120·143
15	150·000	21·428	128·572
15½	160·167	22·881	137·286
16	170·667	24·381	146·286
16½	181·500	25·928	155·572
17	192·667	27·524	165·143
17½	204·167	29·167	175·000
18	216·000	30·857	185·143
18½	228·167	32·581	195·586
19	240·667	34·381	206·286
19½	253·500	36·214	217·286
20	266·667	38·075	228·592
21	294·000	42·000	252·000
22	322·667	46·095	276·572
23	352·667	50·381	302·286
24	384·000	54·857	329·143
25	416·667	58·095	358·572
26	450·667	64·381	386·286
27	486·000	69·428	416·572
28	522·667	74·667	448·000
29	560·667	80·095	480·572
30	600·000	85·714	514·286

No. 4.
Offsets or Ordinates at the end of the first Chain from tangent point of Railway Curves.

Radius of Curve in Chains.	Offsets in Inches and Decimals.	Radius of Curve in Chains.	Offsets in Inches and Decimals.	Radius of Curve in Chains.	Offsets in Inches and Decimals.	Radius of Curve in Chains.	Offsets in Inches and Decimals.
15	26·4000	65	6·0923	125	3·1680	225	1·7600
16	24·7500	66	6·0000	126	3·1428	230	1·7217
17	23·2941	67	5·9104	128	3·0937	235	1·6851
18	22·0000	68	5·8235	130	3·0461	240	1·6500
19	20·8421	69	5·7391	132	3·0000	245	1·6163
20	19·8000	70	5·6571	134	2·9552	250	1·5840
21	18·8571	71	5·5774	135	2·9333	255	1·5529
22	18·0000	72	5·5000	136	2·9117	260	1·5231
23	17·2174	73	5·4246	138	2·8645	265	1·4943
24	16·5000	74	5·3513	140	2·8285	270	1·4667
25	15·8400	75	5·2800	142	2·7887	275	1·4400
26	15·2307	76	5·2105	144	2·7500	280	1·4143
27	14·6666	77	5·1428	145	2·7310	285	1·3995
28	14·1428	78	5·0769	146	2·7123	290	1·3655
29	13·6551	79	5·0126	148	2·6756	295	1·3423
30	13·2000	80	4·9500	150	2·6400	300	1·3200
31	12·7742	81	4·8889	152	2·6052	305	1·2983
32	12·3750	82	4·8292	154	2·5714	310	1·2774
33	12·0000	83	4·7711	155	2·5548	315	1·2571
34	11·6470	84	4·7143	156	2·5384	320	1·2375
35	11·3142	85	4·6588	158	2·5063	325	1·2184
36	11·0000	86	4·6046	160	2·4750	330	1·2000
37	10·7026	87	4·5517	162	2·4444	335	1·1821
38	10·4210	88	4·5000	164	2·4146	340	1·1647
39	10·1538	89	4·4496	165	2·4000	345	1·1478
40	9·9000	90	4·4000	166	2·3855	350	1·1314
41	9·6588	91	4·3516	168	2·3571	355	1·1155
42	9·4285	92	4·3043	170	2·8294	360	1·1000
43	9·2093	93	4·2581	172	2·3023	365	1·0809
44	9·0000	94	4·2128	174	2·2758	370	1·0703
45	8·8000	95	4·1684	175	2·2628	375	1·0560
46	8·6087	96	4·1250	176	2·2500	380	1·0421
47	8·4255	97	4·0825	178	2·2248	385	1·0285
48	8·2500	98	4·0408	180	2·2000	390	1·0154
49	8·0816	99	4·0000	182	2·1758	395	1·0025
50	7·9200	100	3·9600	184	2·1521	400	·9900
51	7·7647	102	3·8824	185	2·1405	410	·9659
52	7·6154	104	3·8077	186	2·1290	420	·9428
53	7·4717	105	3·7714	188	2·1064	430	·9209
54	7·3333	106	3·7358	190	2·0842	440	·9000
55	7·2000	108	3·6667	192	2·0625	450	·8800
56	7·0714	110	3·6000	194	2·0412	460	·8609
57	6·9473	112	3·5352	195	2·0307	470	·8425
58	6·8276	114	3·4736	196	2·0204	480	·8250
59	6·7118	115	3·4435	198	2·0000	490	·8081
60	6·6000	116	3·4138	200	1·9800	500	·7920
61	6·4918	118	3·3559	205	1·9317	510	·7765
62	6·3871	120	3·3000	210	1·8857	520	·7615
63	6·2857	122	3·2459	215	1·8418	530	·7472
64	6·1875	124	3·1935	220	1·8000	540	·7333

No. 5.
Horizontal Distances to an Unit's Height to the following Angles of Elevation.

Angle of Elevation.	Hor. Distance to one Perpendicular.
0° 5′	688
0 10	344
0 15	229
0 30	115
0 45	76
1 0	57
1 15	46
1 30	39
1 45	33
2 0	28
2 15	25
2 30	23
2 45	21
3 0	19
3 15	18
3 28	17
3 35	16
3 49	15
4 6	14
4 24	13
4 45	12
5 12	11
5 42	10
6 21	9
7 7	8
8 8	7
9 27	6
11 19	5
14 2	4
18 26	3
26 34	2
45 0	1

Spotted something that isn't quite right?

Let us know and we'll fix it for future reprints.
And as a thank you for your diligence, we'll
send you a free Sothis book of your choice.

publisher@sothis.press

This page intentionally left blank

Made in United States
North Haven, CT
04 November 2026

21730350R00042